OXFORD LOGIC GUIDES: 13

OXFORD LOGIC GUIDES

1. Jane Bridge: *Beginning model theory: the completeness theorem and some consequences*
2. Michael Dummett: *Elements of intuitionism*
3. A. S. Troelstra: *Choice sequences: a chapter of intuitionistic mathematics*
4. J. L. Bell: *Boolean-valued models and independence proofs in set theory* (*1st edition*)
5. Krister Segerberg: *Classical propositional operators*
6. G. C. Smith: *The Boole–De Morgan correspondence 1842–1864*
7. Alec Fisher: *Formal number theory and computability: a work book*
8. Anand Pillay: *An introduction to stability theory*
9. H. E. Rose: *Subrecursion: functions and hierarchies*
10. Michael Hallett: *Cantorian set theory and limitation of size*
11. R. Mansfield and G. Weitkamp: *Recursive aspects of descriptive set theory*
12. J. L. Bell: *Boolean-valued models and independence proofs in set theory* (*2nd edition*)
13. Melvin Fitting: *Computability theory, semantics, and logic programming*

COMPUTABILITY THEORY, SEMANTICS, AND LOGIC PROGRAMMING

MELVIN FITTING

Lehman College
and
The Graduate School and University Center
City University of New York

OXFORD UNIVERSITY PRESS • NEW YORK
CLARENDON PRESS • OXFORD
1987

Oxford University Press

Oxford London New York Toronto
Delhi Bombay Calcutta Madras Karachi
Petaling Jaya Singapore Hong Kong Tokyo
Nairobi Dar es Salaam Cape Town
Melbourne Auckland

and associated companies in
Beirut Berlin Ibadan Nicosia

Published by Oxford University Press, Inc.,
200 Madison Avenue, New York, New York 10016

Library of Congress Cataloging-in-Publication Data
Fitting, Melvin Chris.
Computability theory, semantics, and logic programming.
(Oxford logic guides; 13)
Includes index.
1. Computable functions. 2. Data structures
(Computer science) I. Title. II. Series.
QA9.59.F58 1986 001.64'2 85-5003
ISBN 0-19-503691-3

2 4 6 8 9 7 5 3 1

Printed in the United States of America
on acid-free paper

PREFACE

To the Student: Computability theory is the study of the limitations and abilities of computers in principle. A mathematical model of a computer is investigated, with no restrictions on memory size or running time (other than finiteness). One establishes the properties of such an idealized machine by writing programs in its "machine language," and by proving results about programs (including results about their very existence or nonexistence).

Of course, to carry on at the machine language level gets very tedious. Just as with real computers, one is quickly led to the introduction of higher and higher level languages (which compile into machine language). In fact, today most people studying computer programming begin with a high-level language and don't see machine language until later. We follow the same pattern here. Early chapters are devoted to the investigation of the theoretical properties of a certain high-level language, which we call EFS. It is not until Chapter 5, when a thorough familiarity with EFS and its properties has been developed, that we introduce our model computer and show how to translate EFS programs into its machine language. It is here we show that working with EFS gives us no more and no less power than we have at the machine language level. What it provides us with is greater intelligibility and convenience.

Because we are beginning with a high-level language for reasons that are pertinent to human understanding, we have further chosen to begin with a language requiring of us minimum concern for mechanical details. We have selected as primary a *logic programming* language. In more conventional programming languages the specification of *what* a program is to do (its logic) is intertwined with the specification of *how* it is to do it (its control). In logic programming languages these aspects are separated. The language EFS that we use is one that allows us to concentrate on a program's logic exclusively. As we will see in Chapter 1, an EFS program simply amounts to a characterization of what will be accepted as output. The mechanism by which that output is to be produced is left entirely up to the language implementation. In Chapter 5 we prove EFS can be correctly implemented (on our mathematical model of a computer). Until then we concentrate on program logic, and ignore control issues entirely.

Because logic programming is a relatively new area, everything we need is developed in the early chapters of this book. Indeed, an understanding of the treatment here will make it comparatively easy to experiment with logic programming on a real computer. The language PROLOG is growing in popularity and is similar to our theoretical language EFS.

What are the major results of computability theory that must be established in a beginning treatment? It is primarily that our notions of computation are *robust* but have essential *limitations* that can not be avoided. Let us elaborate on this statement.

First, our notions of computation are robust. We may say what our subject is in terms of a mathematically defined model computer, but the details of that model are not absolute; we could have created a different model. Or we could have worked with a high-level language quite unlike EFS. Or we could have presented some completely different characterization of computation that does not mention computers or computer languages at all. But it turns out that such approaches have all been proved equivalent. What we are studying is independent of the formalism we introduce to study it. As an analogy, there is a notion of number that is independent of whether we use Arabic or Roman numerals. There is a notion of computation that is independent of computer and computer language. This is a simplified version of the famous Church–Turing Thesis.

We cannot establish the equivalence, here, of all the variety of approaches to computability that have been invented. There are too many of them. We do consider a few important representatives, however. The language EFS is a high-level, nondeterministic, logic programming language. In Chapter 5 we introduce a more conventional, deterministic, imperative-style language. We also introduce a language based on logical notation, a language of a sort appropriate for work with relational data bases. Finally, we introduce a simple computer model and a machine language for it. And we prove the equivalence of all of these by showing how to translate a program in any one of the languages into a program in the others. The same notion of computing underlies these very different characterizations.

We also take up a robustness issue not often treated in computability theory books: insensitivity to change of data structure. We introduce, in Chapter 2, a whole family of EFS-type languages, one for each choice of "built-in" data structure. We discuss the implementability of one data structure in another and its effect on computations. And we show that many common choices of data structures (numbers, words, binary trees, and sets) give rise to languages of equivalent power.

The second major theme is that of essential limitation. We show that there are mathematically well-defined problems whose solutions are not computable, and this is inherent in the nature of computation. The oldest and most famous example is due to Alan Turing and dates from 1936. Computer programs are instructions, to be sure, but they can also be used as data for other programs. Indeed, a compiler is a program whose input and output consists of programs. Turing's Theorem says: It is not possible to write a program which, when fed other programs as data, can correctly predict which of those programs would halt and which would not, when executed. This unsolvability of the halting problem is one of a family of fundamental results that we investigate in Chapter 6.

There is one other major theme we develop, but one not usually found in computability theory books: program semantics. We do this for the family of EFS languages, but not for other language styles. EFS, because of its virtual disregard of control structure, makes the development of a denotational semantics an especially simple task. The semantics that results is one that is familiar in the area of logic programming. But again, background knowledge is not necessary. What we need, we present. This semantics provides us with machinery to prove things about EFS programs with minimal difficulty. Among other things, we use it in a proof of the so-called First Recursion Theorem in Chapter 3, a result of fundamental importance.

In Chapter 3, there is an extended discussion of monotone functions and fixpoints. Similar notions come up in the semantics for other styles of programming languages. We believe they are at their simplest in the logic programming setting we have chosen. After reading this book it should be easier to study program semantics for more conventional languages.

To the Instructor: We give a chapter-by-chapter summary of a more technical nature.

Chapter 1. A simple and natural language called EFS is defined for manipulating strings over an alphabet L. The language is PROLOG-like, and derives from the elementary formal systems of R. Smullyan. Procedures in this language nondeterministically generate sets or relations. In recursive function theory terminology, these are the recursively enumerable sets and relations. Operational and denotational semantics for EFS are defined, and applications to EFS procedures are rigorously developed.

Chapter 2. For each algebraic data structure **S** a programming language EFS(**S**) is defined, and general properties are considered. Specific choices are investigated in detail: numbers, words, trees, sets, and also a miniature relational data base. And a denotational semantics is developed that applies to EFS languages no matter what the choice of data structure. Closure properties of relations generated by procedures are considered. A logical notation is introduced in which such closure properties can be easily stated.

Chapter 3. The definition of EFS procedure is extended so that procedures for *operators* can be written. These are what recursion theorists call enumeration operators. Denotational semantics notions are developed for them, and used to establish their general behavior. Among other things, the Kleene First Recursion Theorem is proved, still allowing for an arbitrary choice of data structure.

Chapter 4. The implementation of one data structure in another is discussed. This can be thought of as a generalization of Gödel numbering. Results are proved that say, roughly, any choice of the data structure from among those of numbers, words, binary trees, or sets is as versatile as any other. The data structure of words with concatenation is chosen as primary for the rest of the book.

Chapter 5. A mathematical model of a computer is presented, based on the *register machines* of Shepherdson and Sturgis. The machine language of our model computer is shown to be equivalent in power to a conventional style imperative language, IMP. The most significant feature of IMP for us is that it is deterministic. The equivalence of EFS and IMP languages is proved. As a tool for doing this, the language LOG is introduced, based on the logical notation presented in Chapter 2. Thus, all of EFS, IMP, LOG, and the machine language of our computer model are shown to be intertranslatable. As a byproduct of this work we have a *structured programming* result: All programming can be done in IMP in a structured manner, and indeed, at most one while loop will be needed. Theorems like this, which have been proved for flowchart programs by Böhm and Jacopini in 1966 and by others for other languages, form the theoretical basis for the current emphasis on structured programming. The result, considered abstractly, is actually an old one. It is the Kleene normal form theorem, from the mid 1930s.

The translation given from EFS to register machine language introduces an obvious exponential blowup in computational time. This allows a discussion of the $P = NP$ and $NP = \text{co-}NP$ problems. Apart from this, computational complexity issues are not considered.

Chapter 6. An interpreter for EFS in itself is written. The existence of a recursively enumerable but not recursive set then follows easily, as does the unsolvability of the halting problem for IMP, and the lack of closure of the LOG definable relations under universal quantification. The Kleene Second Recursion Theorem is proved and discussed. Rice's Theorem and some relatives are proved and applied.

Some 200 exercises are scattered throughout the text. Each chapter ends with a background section giving the origins of the main results presented in that chapter. There is an appendix discussing proofs by induction. Induction is used heavily throughout the book.

Traditional computability theory material is covered, often from a distinctly modern point of view, and several topics are considered that are not included in older computability texts. The level of the material is suitable for an undergraduate senior or beginning graduate computer science course. Either one or two semesters could be allotted, depending on circumstances and the desired depth of coverage.

* * *

I want to thank Laurie Kirby and Ron Sigal for reading and commenting on early drafts of this book. And I especially want to thank Howard Levi and Ken McAloon for useful and frank discussions concerning the text and organization. Most fundamentally, I want to thank Raymond Smullyan for inventing elementary formal systems in the first place.

Montrose, N.Y. M. F.
September 1985

CONTENTS

COMPUTABILITY THEORY, SEMANTICS, AND LOGIC PROGRAMMING

1

A STRING MANIPULATION LANGUAGE

1.1 LANGUAGE BACKGROUND

The specification of a programming language has several different levels: What do the programs look like? What will a program cause a machine to do internally? What does program execution look like from the outside? In formal terms: What are the *syntax, operational semantics,* and *denotational semantics* of a language?

A program is an object made up of letters from an alphabet, put together according to rules of correct formation. These rules and the alphabet constitute the *syntax* of the language. One way of looking at a program is as a syntactic object, with no inherent meaning attached.

If a programming language has been implemented on some machine, then a program in that language will invoke a certain behavior in that machine. One way of assigning a meaning to programs is to give a formal characterization of what we expect it will cause a (perhaps idealized) machine to do. This is the *operational semantics* of a language. Languages that are syntactically identical can have different operational semantics, just as the same word can have different meanings in different languages.

Suppose we have two different programming language implementations, accepting syntactically identical programs. If we run the same program under the two implementations there is no reason to suppose the internal behavior of the machines will be the same in both cases. In fact, we don't usually care. What is important to us is what we see from the outside. If we supply the same input we should get the same output in each case. This input-output behavior mathematically is a *function.* Specifying functional behavior of programs without concern for "internal workings" constitutes the *denotational semantics* of a language. We think of a program as a way of denoting a function. Usually the denotation of a program is specified in terms of the denotations of subprograms, rather than in machine behavior terms. Programs can have different meanings in the sense of operational semantics, but be identical in the denotational sense. That is, different programs can "cause" a machine to compute the same function, but in different ways. When one asks if two language implementations are equivalent one means, do syntactically identical programs denote the same semantic object, the same function. We do not expect the operational semantics to be the same. Indeed, having distinct operational semantics is what is usually *meant* by having different implementations.

In this chapter, we introduce a language we call *EFS* (standing for elementary formal system). We present its syntax in §1.3, its operational semantics starting in §1.4, and its denotational semantics starting in §1.7. This and closely related languages will be with us throughout the book.

1.2 INFORMALITIES

The language EFS is probably quite unlike programming languages you have experience with. Here we discuss essential differences from an intuitive point of view before getting to formal details starting in §1.3.

First, identifiers will be *multiple valued,* rather than single valued as in most languages. We may speak of *a* value of an identifier rather than *the* value. A better way to think of it is: identifiers have *sets* as values. For example, when working with numbers, we will not have an identifier that successively takes on the even numbers as values, one after the other. But we will have an identifier that acquires *all* the even numbers as values. *A* value of such an identifier is 2, and another is 4, and so on. *The* value of it is the *set* of even numbers.

Assignment statements will be somewhat different as well. In most programming languages they are also "disassignment" statements; that is, assigning a value to an identifier removes the previous value. Here, once an identifier acquires a value, it keeps it. It may acquire more, but it never loses any.

Program execution is *nondeterministic.* You are probably used to language implementations in which, at each stage of a computation, the next step is uniquely determined and is specified implicitly by the program itself. Under a nondeterministic operational semantics we can write programs that "say" to a machine what *may* be done, rather than what *must* be done. Different runs of a program can be different, although none can violate our instructions. An informal example may help here. Suppose we know what the even numbers are. How do we get the odd numbers? Add 1 to the even numbers. This means that every time we arbitrarily pick an even number, and add 1 to it, we get an odd number; and we can get any particular odd number by making a suitable choice of even number and adding 1 to it. We have characterized precisely the odd numbers, in terms of the even numbers, but our characterization allows for many different calculations to be performed.

Nondeterminism is reasonable given our choice of identifier behavior. Because our identifiers can have many values, with perhaps no natural order to them, which do we use when a value is needed in a computation? We assume the computer on which the language is implemented is capable of making an arbitrary choice, one which may or may not be the same each time.

In Chapter 5, we show the apparent extra power that nondeterminism adds is not real, at least in principle. We do this by showing how to translate all EFS programs into a more conventional, fully deterministic language. Until then, all languages considered will have nondeterministic features.

Programming languages generally are tight mixtures of control structures and data structures. For example, allowing counting loops presupposes that the counting numbers are available. But we want the control structures of our languages to be as general as possible because we will be considering many different choices of basic data structures. We take the following small number of control structures as fundamental.

We have assignment statements: Give this (additional) value to an identifier. We have conditional assignment statements: Assign this value provided certain conditions are met. We have procedure calls, which can be recursive, that is, a procedure can call itself. And that's it. Nondeterminism comes in here too: Program instructions are not given an inherent operational semantic order; first do this, next do that, and so on. Order of execution may differ in different runs. It is not ours to specify when writing the programs. Let us consider another informal example.

Suppose, as above, we know what the even numbers are. Which nonnegative numbers are *not* powers of 2? Well, a number is not a power of 2 if it is not 0 and it has an odd factor other than 1. So, consider the following "instructions."

1. If x is an even number, then $x + 1$ is an odd number.
2. If y is an odd number other than 1, and z is any number other than 0, then $y * z$ is not a power of 2.

In following these we could use (1) several times to generate a large set of odd numbers, after which we could use (2) with each of these odd numbers. Or, we could alternate applications of (1) and (2): Every time we determine a number is odd using (1) we could use (2) with it immediately. Any many other patterns are also possible. The point is, our "instructions" will never tell us something is not a power of 2 when it is, no matter how we follow them, and if something is not a power of 2, there is some way of determining that by following the "instructions."

Finally, a refinement on what was said above. The values of identifiers will be sets *and relations*. How do we use the term relation? Consider the (informal) relation of being someone's telephone number. We assume several people can share a number, and some people might have several numbers. In effect, the telephone book defines the relation. The telephone book lists *pairs*, each pair consisting of a name and a telephone number, in that order. This is the model we take for relations in general. We assume that if we have n things, $w_1, \ldots, w_n$, we can form the *ordered n-tuple* $\langle w_1, \ldots, w_n \rangle$ consisting of those n things in a particular order. Then an n-place *relation* is simply a set of ordered n-tuples. For example, the telephone-number-of *relation* could be defined here as the set of all ordered pairs $\langle w_1, w_2 \rangle$ where w_1 is a person's name and w_2 is a telephone number assigned to that person. As another example consider the 3-place relation consisting of ordered triples $\langle w_1, w_2, w_3 \rangle$ where w_1 is a person's name and w_2 and w_3 are names of the person's mother and father. Such a relation embodies some of the information found in genealogy records.

In EFS, identifiers will stand for sets and relations in the sense above. Indeed a set, as we use it, is essentially just a 1-place relation, where we identify the one tuple $\langle x \rangle$ with x itself.

Most computer languages work with *functions* rather than relations. We do not introduce any specific function mechanism here. Instead we work with functions via their *graphs*. For example, in arithmetic, the graph of the function $f(x) = 2x + 1$ is the set of ordered pairs $\langle x, y \rangle$ where $y = f(x) = 2x + 1$. Similarly, the addition function, $+$, has as its graph the 3-place relation consisting of all $\langle x, y, z \rangle$ where $z = x + y$. This use of graphs gives us relations to work with, and nothing is lost to us. If we need to know, for instance, what the sum of 2 and 3 is, we can look through the graph of the addition function until we find a triple whose first two components are 2 and 3 (there will be only one), and take the third component. Thus, we have the ability to add without the need to introduce a specific function mechanism; the one for relations is sufficient.

There are many common places where graphs (relations) are substituted for functions. For example, if one uses a pocket calculator to compute sines or cosines, one has an example of a function in action. A number is entered, and a number is returned. But if one uses a sine or cosine table instead, one is taking the relation approach. To compute the sine of an angle, one looks through the table until the angle is found in the appropriate column, then one takes the number corresponding to it in the adjacent column. The similarity of this to the way we are treating functions should be obvious.

The languages used in this book are, for the most part, *logic programming* languages. Variants have been implemented and used. A widely available language that is close to EFS is *PROLOG*. But PROLOG, for reasons of efficiency, has been given a deterministic operational semantics and often behaves quite differently from a "pure" logic programming language. Mathematically the computer language family used here derives from a construct developed in the 1950s, known as *elementary formal systems*. So we have chosen the name *EFS* language, standing for elementary formal system. In Chapter 5 we will show the equivalence of an EFS language using a character string data structure, and a more conventional kind of programming language. Until then, only EFS languages will be used.

1.3 SYNTAX

In the next chapter we present a whole family of EFS languages, for arbitrary choices of data structures. So that you may gain experience before generality, this chapter presents a representative example: an EFS language for a data structure of character strings over an alphabet L. Since nothing critical depends on the choice of alphabet we do not fix it once and for all, but leave its specification open. This section, then, presents the syntax of a language we call EFS(**str**(L)).

Definition. By an *alphabet* we mean a finite nonempty set of objects called *letters* or *symbols*. If L is an alphabet, by L^* we mean the collection of all finite sequences of members of L, called *words* or *strings* over L. We allow the *empty* or *null* word, with no letters and denote it by Δ.

For example, $\{a, b, c\}$ and $\{0, 1\}$ are alphabets. '*abcba*' is a 5-letter word over the alphabet $\{a, b, c\}$. '101101' is a 6-letter word over the alphabet $\{0, 1\}$. As is common with programming languages, we have used quotes to indicate beginning and ending of words. This is necessary to avoid the confusion that can arise because the space is often an allowed letter. We will modify this convention later on. Typographically we do not distinguish between a 1-letter word and its only letter. In practice this will cause no confusion because the only entities our EFS language manipulates are words, never letters as such. There is no length limit on words (we place no memory restrictions on our idealized computers), so L^* has infinitely many members.

Words can be *concatenated* to form new words. For example, '*abc*' and '*bac*' concatenate (in this order) to produce '*abcbac*'. Concatenation is a function, but recall we do not have a mechanism for dealing with functions as such in EFS languages. We talk about the 3-place concatenation *relation* instead.

Definition. $\mathbf{CON}_L$ is the 3-place relation consisting of all ordered triples $\langle w_1, w_2, w_3 \rangle$ where each of w_1, w_2, and w_3 are words over L, and w_3 is the result of concatenating w_1 and w_2.

For example, the triples $\langle$'*abc*', '*bac*', '*abcbac*'$\rangle$, $\langle$'*bac*', '*abc*', '*bacabc*'$\rangle$, and $\langle$'*abc*', Δ, '*abc*'$\rangle$ are all members of $\mathbf{CON}_L$, assuming L includes the letters a, b, and c. The relation $\mathbf{CON}_L$ depends on the choice of L; we don't concatenate words we can't make. When no confusion is likely, though, we will often write **CON** to simplify notation.

Definition. $\mathbf{str}(L)$ is the *data structure* $\langle L^*, \mathbf{CON}_L \rangle$, where L is an alphabet. The EFS language for this data structure is denoted EFS($\mathbf{str}(L)$).

Thus, in $\mathbf{str}(L)$, the objects are the words in L^*, and the things we "know" about them are the members of $\mathbf{CON}_L$, that is, we know about concatenation. Now we continue with the specification of EFS($\mathbf{str}(L)$) syntax.

Definition. An *identifier* is any string made up of letters from "standard" alphabet, A, B, . . . , Z of capital letters together with the "break" symbol, '_', and that does not begin or end with '_'.

For example, ODD, EVEN, CON, and PART_OF are identifiers.

For reasons that will be clear in the next section, in addition to words of L^* and identifiers, we also need an extra, unlimited supply of things called *variables*. These behave as place-holders in the stating of procedure instructions, and do not play a direct role in computations.

Definition. The *variables* are $v, v', v'', v''', \ldots$. Informally, we generally write x, y, z, etc. to keep things readable. The only restriction is that neither $'$ nor v should be in L, the alphabet we are given to compute over. We want to able to distinguish variables from constants.

Definition. The remainder of our language basics consists of the following *punctuation* symbols.

arrow	$\rightarrow$
parentheses	)(
comma	,

That's it.

Note. To keep the notation uncluttered, from now on we make the following convention. We assume the alphabet L does not contain any of the punctuation symbols, space, the letters A, B, . . . , Z, _, or v or '. That is, it is disjoint from symbols playing a special role in EFS. With this convention it is no longer necessary to use quotes to delineate members of L^* since no confusion can arise. Then, from now on, we write *abc* rather than '*abc*'.

Definition. The words of L^*, and the variables, are *terms* of EFS(**str**(L)). Nothing else is a term.

We use prefix relation notation. Let IDENT be an identifier and $t_1, \ldots, t_n$ be terms. IDENT $(t_1, \ldots, t_n)$ can be read informally as: the n-tuple $\langle t_1, \ldots, t_n \rangle$ is in the relation (named by) IDENT. Thus, if CON designates the concatenation relation, CON(*ab, bc, abbc*) is informally true.

Definition. If $t_1, \ldots, t_n$ are terms, and IDENT is any identifier, then IDENT$(t_1, \ldots, t_n)$ is an *atomic statement.*

For example, PART_OF(*bc, abcd*) and PLUS(11, x, y) are atomic statements (assuming L has a, b, c, d, 1 as letters, and x and y are variables). There will be no prohibition against an identifier representing a set of n-tuples and also a set of k-tuples in the same procedure. The two roles can be differentiated by simply counting the number of terms that follow a given occurrence of an identifier. In practice, of course, it makes procedures harder to read and should be avoided.

Definition. If $S_1, S_2, \ldots, S_n$ is a list of atomic statements, and T is an atomic statement, then $S_1 \rightarrow S_2 \rightarrow \ldots \rightarrow S_n \rightarrow T$ is a *statement.* We allow the list $S_1, S_2, \ldots, S_n$ to be empty. That is, if T is an atomic statement, then T itself is a statement too.

For example, PART_OF(*bc, abcd*) and PART_OF(x, y)$\rightarrow$PART_OF(y, z)$\rightarrow$PART_OF(x, z) are statements. In $S_1 \rightarrow S_2 \rightarrow \ldots \rightarrow S_n \rightarrow T$, the T

and the S_i are intended to play different roles. T is an *assignment,* $S_1, S_2, \ldots, S_n$ are *conditions* or *tests.* For example, the intended behavior of $\mathrm{A}(w) \to \mathrm{B}(w, z) \to \mathrm{D}(z)$ will be as follows: z should be added to the set of values of D provided w is one of the values of A and $\langle w, z \rangle$ is one of the values of B. If there are no conditions, we have an *unconditional* assignment statement.

Definition. For an atomic statement T, we say T is in the *assignment position* in the statement $S_1 \to S_2 \to \ldots \to S_n \to T$, and also in the unconditional statement T itself.

In "running" a procedure, certain relations are being computed, or generated, while others are assumed known, or given. For example, when using the structure **str**(L) we will assume the concatenation relation $\mathbf{CON}_L$ is known. Relations that are given to us to work with will be represented by identifiers but *we* should not assign values to those identifiers. Since the collection of relations we are given as primitives will change from time to time, we do not set up a special notation for it. Instead, we simply say what identifiers are considered reserved, representing what "given" relations. We use the term *work space* to mean, loosely, what is known, what is available for use. Procedures are written "in" some work space.

Definition. A *work space* **W** consists of the following three items. (1) A specification of a *domain,* the objects we are talking about. (2) A list of identifiers, designated as *reserved.* (3) A specification of what relations the reserved identifiers represent. We call the relations represented by the reserved identifiers the *given* relations of **W**.

The *basic work space* of EFS(**str**(L)) is the following. (1) the domain is L^*, all words over the alphabet L. (2) the only reserved identifier is CON. (3) CON represents the concatenation relation on L, $\mathbf{CON}_L$.

Do not confuse CON and $\mathbf{CON}_L$. CON is an identifier, three letters long. $\mathbf{CON}_L$ is the concatenation relation itself. Part of the work space specification is that CON represents $\mathbf{CON}_L$. Without this specification, there need be no connection between them.

Definition. We call a procedure statement *acceptable* (in a work space) if no reserved identifier occurs in the assignment position.

For example, if D is not a reserved identifier, then $\mathrm{A}(w) \to \mathrm{B}(w, z) \to \mathrm{D}(z)$ is an acceptable statement. Thus, a statement is acceptable if it does not "try" to assign values to a reserved identifier. Our general practice, much like in LISP or LOGO or FORTH, is to start off with a basic work space, then to enhance it. In the work space we start with, the relations represented by reserved identifiers can be assumed to be "given." We don't care about the mechanism by which this is done when we write our procedures. In practice, of course, some machine language specification would be involved. Then, starting with the basic work space, we write procedure after procedure, each one characterizing some additional set or relation (by means to be given in the next section). Whenever we have thus characterized a new relation, we can designate an identifier to represent it, as a new reserved identifier. The procedure we have written constitutes the specification of what this identifier represents. In this way we produce a richer work space.

Definition. Suppose we are given a work space **W**. A *procedure in* **W** consists of two parts: a header and a body.

The *header* of a procedure designates an unreserved identifier and a number. The identifier can be thought of as the name of the procedure, and also as the identifier that represents that procedure's output. The number specifies how many places the output relation has: n means the output is an n-place relation. The format we use for the header is this.

NAME(n):

Here NAME is an identifier and n is a number.

The *body* of a procedure is any finite collection of statements that are acceptable in the work space **W**. The format we use for a procedure body is simply to list the procedure statements involved, separated by semicolons, and terminated with a period, as in Pascal.

Finally, in the procedure with header NAME(n):, NAME is the *output identifier*. All identifiers occurring in the procedure body other than NAME and reserved identifiers are *internal identifiers* of the procedure. (NAME can also be an internal identifier if it is used as a k-place identifier, where $k \neq n$.)

For example, suppose L is the alphabet $\{0, 1\}$, and the work space is the basic work space of EFS(**str**(L)). Then the domain consists of all words over the alphabet $\{0, 1\}$; the only reserved identifier is CON; and CON represents the concatenation relation on words over $\{0, 1\}$. The following is a procedure in this work space. (Recall that Δ denotes the empty word.)

SAMELENGTH (2):
    SAMELENGTH(Δ, Δ);
    SAMELENGTH(0, 1);
    SAMELENGTH(1, 0);
    SAMELENGTH(0, 0);
    SAMELENGTH(1, 1);
    SAMELENGTH(x, y)$\rightarrow$SAMELENGTH(z, w)$\rightarrow$
        CON(x, z, u)$\rightarrow$CON(y, w, v)$\rightarrow$SAMELENGTH(u, v).

We have broken this procedure up into separate lines, added spaces, and used indentation to enhance readability. Spaces and new lines play no official role, however. In this procedure, SAMELENGTH is the output identifier and there are no internal identifiers.

Recall that there was no prohibition against an identifier being used to represent both an n-place and a k-place relation in the same procedure. Consequently, it is not enough to say, in a procedure header, which identifier represents output. We must also say the *arity* of the relation, to ensure against possible confusion.

We have now completed the description of the syntax of the language EFS(**str**(L)). In the next section we will say what a computation is. Then it will be seen that, using the procedure given above, SAMELENGTH represents the relation that holds between two words over $\{0, 1\}$ just when the words have the same lengths. In the next section, we give more examples of procedures.

Exercise 1.3.1. Which of the following are, and which are not, procedures in the basic work space of EFS(**str**(L))?

(a) PARTA (2):
$\quad$ CON(x, y, z)→PARTA(x, z);
$\quad$ PARTA(x, y)→PARTA(y, z)→PARTA(x, z).

(b) PARTB (1):
$\quad$ PARTB(Δ).

(c) PARTC (2):
$\quad$ CON(x, y, z)→PARTC(y, z);
$\quad$ PARTC(Δ, x)→CON(x, y, x).

(d) PARTD (2):
$\quad$ CON(x, y, z)→PARTD(y, z);
$\quad$ PARTD(x, y)→PARTA(y, x).

(e) PARTE (2):
$\quad$ PARTA(x, y)→PARTE(y, x).

Exercise 1.3.2. Suppose PARTA is added to the list of reserved identifiers of Exercise 1.3.1. Then which of (a) through (e) are procedures?

1.4 OPERATIONAL SEMANTICS

We describe EFS(**str**(L)) *computations*. Any implementation of the language would have to be capable of carrying out the steps of such a computation, through each step would, itself, be broken down into a large number of simple machine operations. We do not go below the level of computations; it is what is relevant here. In what follows, we are making certain assumptions about the idealized computer on which EFS(**str**(L)) is being implemented. We assume the members of L^* have some machine representation (no matter how long they may be). We assume that if a particular triple $\langle w_1, w_2, w_3\rangle$ is in $\mathbf{CON}_L$ the computer has a means of determining it, that is, it has some mechanism for systematically generating a "concatenation table." Note, we did not say that if $\langle w_1, w_2, w_3 \rangle$ is not in $\mathbf{CON}_L$, the computer can determine that fact. We assume the $\mathbf{CON}_L$ table itself can be generated; that is all. The details of the representation of members of L^*, and the mechanism for generating (finite portions of) $\mathbf{CON}_L$ do not concern us. That is below the level at which we are describing things. We assume it has been done, and we go on from there.

Definition. Recall that procedure statements were allowed to contain variables. By a *substitution instance* of a procedure statement S over L^* we mean the result of replacing, in S, every variable by a word in L^*, provided that every occurrence of the same variable is replaced by the same word.

For example, suppose L is the alphabet $\{a, b\}$, and consider the statement S: $\mathrm{D}(x) \rightarrow \mathrm{E}(y, ab) \rightarrow \mathrm{F}(ab, y, x)$ where x and y are variables. Then $\mathrm{D}(bb) \rightarrow \mathrm{E}(aba, ab) \rightarrow \mathrm{F}(ab, aba, bb)$ is a substitution instance of S, where we have replaced x by bb and y by aba. Another substitution instance is $\mathrm{D}(\Delta) \rightarrow \mathrm{E}(a, ab) \rightarrow \mathrm{F}(ab, a, \Delta)$. Thus, we think of a statement with variables as standing for a whole family of statements without variables, all having the same structure.

Definition. Say we have a work space $\mathbf{W}$, and we have a procedure P in $\mathbf{W}$. A *computation* for P in $\mathbf{W}$ goes in stages. At each stage of a computation for P:

1. Select one of the procedure statements of P.
2. Select a substitution instance of that statement, say it looks like $X_1 \rightarrow X_2 \rightarrow \ldots \rightarrow X_p \rightarrow \mathrm{IDENTIFIER}(t_1, \ldots, t_j)$.

and

3. If each of the *conditions* $X_1, X_2, \ldots, X_p$ has been previously *verified*, assign $\langle t_1, \ldots, t_j \rangle$ as a value of IDENTIFIER. (If the procedure statement chosen from P has no conditions, the assignment is always executed.)

A condition $R(u_1, \ldots, u_k)$ has been *verified* if: R is a reserved identifier of $\mathbf{W}$ and $\langle u_1, \ldots, u_k \rangle$ is in the relation $\mathbf{W}$ associates with R, or R is unreserved and at an earlier stage of the computation $\langle u_1, \ldots, u_k \rangle$ was assigned to R.

There are two places in each computation stage where a choice must be made: in selecting which procedure statement to work with, and in selecting which substitution instance to use. This is what makes computations nondeterministic. We will see in Chapter 5 how procedures in this language can be translated into a language without nondeterministic features.

Definition. Again, let P be a procedure in work space $\mathbf{W}$, and let IDENTIFIER be an identifier. We say $\langle t_1, \ldots, t_j \rangle$ is in the relation that IDENTIFIER *represents* using P if there is at least one computation for P in $\mathbf{W}$ assigning

$\langle t_1, \ldots, t_j \rangle$ as a value of IDENTIFIER (or if IDENTIFIER is reserved and $\langle t_1, \ldots, t_j \rangle$ is one of its given values). The *output* of a procedure in **W** with header NAME(n): is simply the set of n-tuples that are in the relation NAME represents using that procedure.

For example, consider the procedure SAMELENGTH given in the previous section, and which we repeat here for convenience.

SAMELENGTH (2):
  SAMELENGTH(Δ, Δ);
  SAMELENGTH(0, 1);
  SAMELENGTH(1, 0);
  SAMELENGTH(0, 0);
  SAMELENGTH(1, 1);
  SAMELENGTH(x, y)$\rightarrow$SAMELENGTH(z, w)$\rightarrow$
    CON(x, z, u)$\rightarrow$CON(y, w, v)$\rightarrow$SAMELENGTH(u, v).

W is the basic work space of EFS(**str**(L)): the domain is L^* where $L = \{0, 1\}$; the only reserved identifier is CON; CON represents the concatenation relation on L^*. We show that $\langle 01, 10 \rangle$ is in the output of this procedure in the work space **W**. That is, we show there is a computation that assigns $\langle 01, 10 \rangle$ to SAMELENGTH. We describe the stages of one such computation.

Stage 1: SAMELENGTH(0, 1) is a substitution instance (trivially) of the second procedure statement: $\langle 0, 1 \rangle$ is assigned as a value of SAMELENGTH.

Stage 2: SAMELENGTH(1, 0) is a substitution instance of the third procedure statement: $\langle 1, 0 \rangle$ is assigned as a value of SAMELENGTH.

Stage 3: SAMELENGTH(0, 1)$\rightarrow$SAMELENGTH(1, 0)$\rightarrow$CON(0, 1, 01)$\rightarrow$CON(1, 0, 10)$\rightarrow$SAMELENGTH(01, 10) is a substitution instance of the last procedure statement. The conditions of this have been verified, so $\langle 01, 10 \rangle$ is assigned as a value to SAMELENGTH.

It is easy to come up with computations using SAMELENGTH that do *not* assign $\langle 01, 10 \rangle$ to SAMELENGTH. Just make "stupid" choices every time some choice of procedure statement or substitution instance is to be made. The definition said that $\langle 01, 10 \rangle$ would be in the output if *some* computation assigned $\langle 01, 10 \rangle$ to SAMELENGTH, not if *every* computation did so. A computation that does not assign a certain value to the output identifier is simply a computation of something else. Showing something is not an output of a given procedure is a much subtler business.

Exercise 1.4.1.

(a) Give a procedure, to be called SHORTER, in the work space used above, whose output should be those pairs $\langle t_1, t_2 \rangle$ where t_1 is a shorter word than t_2.

(b) Use your SHORTER procedure and show that $\langle 01, 101\rangle$ really is an output.

Next we want to strip the description of a computation to its essentials. The following sequence of statements displays the key items used in the example above. (The numbers at the left are only for reference purposes.)

(1) SAMELENGTH(0, 1)
(2) SAMELENGTH(1, 0)
(3) CON(0, 1, 01)
(4) CON(1, 0, 10)
(5) SAMELENGTH(0, 1)→SAMELENGTH(1, 0)→
CON(0, 1, 01)→CON(1, 0, 10)→SAMELENGTH(01, 10)
(6) SAMELENGTH(01, 10)

In this, (3) and (4) are given instances of CON; (1), (2), and (5) are substitution instances of procedure statements; and (6) "follows from" lines (1) through (5) by the mechanism we introduced for conditional assignments.

Any computation will give rise to a similar sequence of statements. And it is easy to characterize those sequences of statements that arise from computations. Each line must be either (1) a substitution instance of a procedure statement, or (2) an instance of a given relation of the work space, or (3) must be T, where an earlier line is $X_1 \rightarrow X_2 \rightarrow \ldots \rightarrow X_p \rightarrow T$, and still earlier lines are $X_1, X_2, \ldots, X_p$ (where $X_1, X_2, \ldots, X_p$, and T are all atomic).

The statements that occur in such a sequence because of conditions (1) and (2) can actually occur many times. A substitution instance of a procedure statement remains one no matter how many times we list it. This allows us to relax the third condition somewhat. $X_1 \rightarrow \ldots \rightarrow X_p \rightarrow T$ can occur in the sequence before some of $X_1, \ldots, X_p$, and we can still conclude T. If $X_1 \rightarrow \ldots \rightarrow X_p \rightarrow T$ occurs too soon, it could be repeated later, and so allow us to conclude T by condition (3) as stated above. Since this is always the case, we will dispense with the actual repetition of $X_1 \rightarrow \ldots \rightarrow X_p \rightarrow T$. This is a minor point, but it tends to make things simpler in practice. We officially record this version of condition (3) under the name:

Assignment Rule. From $X_1, X_2, \ldots, X_p$, and $X_1 \rightarrow X_2 \rightarrow \ldots \rightarrow X_p \rightarrow T$ (in any order) we may conclude T. (It is assumed that all the X's and T are atomic statements without variables.)

Definition. We call a sequence of statements a *trace,* from a procedure P, in a work space **W**, if, for each statement in the sequence, either:

1. It is a substitution instance of a procedure statement of P, or
2. It is an instance of a given relation of the work space **W**, or
3. It follows from earlier statements by the Assignment Rule.

From now on the display of a trace will be taken as sufficient evidence for the existence of a computation, and computations themselves will rarely be discussed.

A trace whose last line is IDENTIFIER($t_1, \ldots, t_j$) establishes that $\langle t_1, \ldots, t_j \rangle$ is in the relation that IDENTIFIER represents.

Exercise 1.4.2. Use the SHORTER procedure written for Exercise 1.4.1, and give a trace whose last line is SHORTER(1, 000).

The notion we are calling a trace is familiar to mathematical logicians under the name *derivation*. In the terminology of formal logic, our *axioms* are substitution instances of procedure statements and given instances of reserved identifiers. And the only *rule of inference* is the Assignment Rule stated above. Most inference rules in logic do not have a variable number of premises, as our Assignment Rule does. But, as a matter of fact, the Assignment Rule can be considerably restricted, without changing anything essential.

Restricted Assignment Rule. Suppose X is atomic, but F is not necessarily atomic. From X and $X \rightarrow F$ (in either order) we may conclude F.

For example, if A, B, C, and D are all atomic, then starting with A, B, C and $A \rightarrow B \rightarrow C \rightarrow D$ we could conclude D directly using the original Assignment Rule. But using the Restricted Assignment Rule, from A and $A \rightarrow B \rightarrow C \rightarrow D$ we could conclude $B \rightarrow C \rightarrow D$, and from that and B, we could conclude $C \rightarrow D$, and finally, from that and C, we could still conclude D.

Definition. A sequence of statements is a *restricted trace* if it meets the conditions for being a trace except that the Restricted Assignment Rule is used instead of the Assignment Rule.

It is easy to see that if there is a restricted trace whose last line is IDENTIFIER($t_1, \ldots, t_j$), then there is a trace with the same last line, and conversely. So it does not matter which version we use. The Restricted Assignment Rule will be of use to us in later chapters when we construct an interpreter for EFS(**str**(L)), because it is simpler in its workings. Also, it is a rule of inference that is well known to logicians under the name *modus ponens*. For now, however, it is generally easier to display a trace, rather than a restricted trace, and that is what we will do.

In the following exercises, recall that, for a given alphabet L, the basic work space of EFS(**str**(L)) is the one whose domain is L^* and whose only reserved identifier is CON, which represents concatenation on L^*.

Exercise 1.4.3. Let $L = \{0, 1\}$ and consider the following procedure in the basic work space of EFS(**str**(L)).

SUC (2):
    SUC(1, 10);
    CON(x, 0, y) → CON(x, 1, z) → SUC(y, z);
    SUC(x, y) → CON(x, 1, u) → CON(y, 0, v) → SUC(u, v).

SUC is intended to represent the successor relation in base 2 notation. Give traces to establish that $\langle 0, 1\rangle$, $\langle 1, 10\rangle$, $\langle 10, 11\rangle$, and $\langle 11, 100\rangle$ are all in the output of this procedure.

Can you still do this problem if the procedure statement SUC(1, 10) is replaced by SUC(Δ, 1)?

Exercise 1.4.4. This time let L be the one-letter alphabet $\{1\}$, and use the basic work space of EFS(**str**(L)). Consider the following procedure.

FIB (1):
SEQ(1, 1);
SEQ(x, y)$\rightarrow$CON(x, y, z)$\rightarrow$SEQ(y, z);
SEQ(x, y)$\rightarrow$FIB(x).

FIB is intended to represent the Fibonacci numbers in the following sense. A word consisting of n 1's is supposed to be in the output just when n is a Fibonacci number. (The Fibonacci numbers are the terms in the sequence $1, 1, 2, 3, 5, 8, \ldots$, where the first two terms are ones, and after that each term is the sum of the two terms preceding.)

Show that 11, 111, 11111 are in the output of FIB.

Exercise 1.4.5. Let $L = \{a, b, c\}$ and use the basic work space of EFS(**str**(L)). Give a procedure whose output consists of those words not containing c.

Exercise 1.4.6. Let $L = \{0, 1\}$ and use the basic work space of EFS(**str**(L)). Give a procedure whose output consists of all words of the form 1(string of 0's)1, that is, consists of 11, 101, 1001, 10001, etc.

1.5 EXPANDING A WORK SPACE

So far we have only used basic work spaces. The effect was that the writing of every procedure started from the beginning each time. From now on, we will allow ourselves to make use of the results of previously written procedures; we will take them as given as well.

Definition. Let **W** be a work space. Suppose we write a procedure in this work space, say

NAME (n):
(body of procedure NAME).

The output of this procedure in the work space **W** is an n-place relation. By the *expansion* of **W** by this procedure we mean the work space **W**′ that is like **W** except that it also includes NAME as a reserved identifier, with NAME specified as representing the output of the procedure NAME in the original work space **W**.

For example, begin with the basic work space of EFS(**str**(L)) where $L = \{1\}$. Consider the following procedure in that work space.

EVEN (1):
    EVEN(Δ);
    EVEN(x)→CON(x, 11, y)→EVEN(y).

It is not hard to see that the output of this procedure in the basic work space is the set of all words made up of an even number of 1's.

Now we expand the basic work space by adding EVEN as a reserved identifier, representing the output of the procedure above in the basic work space; that is, it represents the set of even-lengthed words. In the expanded work space we have two reserved identifiers, CON and EVEN. Note that in this expanded work space we can no longer use EVEN in the assignment position of any procedure statement; acceptable statements are relative to a work space. Here is a procedure in the expanded work space.

ODD (1):
    EVEN(x)→CON(x, 1, y)→ODD(y).

Clearly, ODD represents the set of words whose length is odd.

These two simple procedures already illustrate some important features. In the procedure ODD, the conditions in the only procedure statement involve EVEN and CON. So in computations it will be necessary to use information that something is in the set that EVEN represents, and something else is in the set that CON represents. That is, a computation will have at least one *procedure call* on EVEN and at least one on CON. Thus, conditions can be thought of as procedure calls that must be successfully completed before an assignment can be made. The procedure ODD calls on the previously written EVEN, but in the second procedure statement of EVEN, the procedure calls on *itself*. Procedures that do this are often called *recursive*. Recursive procedure calls are fundamental in the EFS languages being presented here. We lack machinery like counting loops; we use recursion instead.

From now on, programming in an EFS language will take the following form. We start with a basic work space. That is expanded by writing a procedure. That in turn may be expanded, and so on, until we reach a work space, and a procedure in it, that does what we want. Generally we will be rather informal about specifying work spaces in detail. We assume that all previously written procedures are available. Once a procedure has been written, we assume that later procedures can make use of it; later procedures are written in a work space expansion using the earlier procedure. We will only rarely give an explicit list of reserved identifiers and procedures to specify what they represent. The following example and exercise should illustrate the process.

Let $L = \{0, 1\}$ and begin with the basic work space of EFS(**str**(L)). We write the following procedure in this work space.

SUC (2):

$\mathrm{SUC}(\Delta, 1)$;
$\mathrm{CON}(x, 0, y) \to \mathrm{CON}(x, 1, z) \to \mathrm{SUC}(y, z)$;
$\mathrm{SUC}(x, y) \to \mathrm{CON}(x, 1, u) \to \mathrm{CON}(y, 0, v) \to \mathrm{SUC}(u, v)$.

(This appeared earlier in Exercise 1.4.3.) In fact, SUC represents the successor relation in base 2 notation. Now expand the work space by adding SUC as a reserved identifier, representing the output of the procedure above. The following procedure is written in this expanded work space.

ADD (3):

$\mathrm{ADD}(x, \Delta, x)$;
$\mathrm{ADD}(\Delta, x, x)$;
$\mathrm{ADD}(x, y, z) \to \mathrm{CON}(x, 0, u) \to \mathrm{CON}(y, 1, v) \to \mathrm{CON}(z, 1, w) \to$
$\quad \mathrm{ADD}(u, v, w)$;
$\mathrm{ADD}(x, y, z) \to \mathrm{CON}(x, 1, u) \to \mathrm{CON}(y, 0, v) \to \mathrm{CON}(z, 1, w) \to$
$\quad \mathrm{ADD}(u, v, w)$;
$\mathrm{ADD}(x, y, z) \to \mathrm{CON}(x, 0, u) \to \mathrm{CON}(y, 0, v) \to \mathrm{CON}(z, 0, w) \to$
$\quad \mathrm{ADD}(u, v, w)$;
$\mathrm{ADD}(x, y, z) \to \mathrm{CON}(x, 1, u) \to \mathrm{CON}(y, 1, v) \to \mathrm{SUC}(z, t) \to$
$\quad \mathrm{CON}(t, 0, w) \to \mathrm{ADD}(u, v, w)$.

In the expanded work space, ADD represents the base 2 addition relation. The first two procedure statements in ADD simply say that if there is nothing to add, nothing changes. The next three statements are intended to take care of addition without carry. Say x, y, and z are strings of base 2 digits, and we already know that $x + y = z$. These procedure statements are used to "calculate" $x0 + y1$, $x1 + y0$, and $x0 + y0$. The last procedure statement covers adding with a carry. Note that it involves a procedure call on SUC.

Exercise 1.5.1. Show the following are in the output of ADD, in the expanded work space.

(a) $\langle 10, 1, 11 \rangle$
(b) $\langle 100, 10, 110 \rangle$
(c) $\langle 11, 1, 100 \rangle$

Now we can expand the work space again, by throwing in ADD as a reserved identifier, with the procedure above to specify what it represents. The reserved identifiers now are CON, SUC, and ADD. In this new work space it is possible to write a procedure for the multiplication relation. We can use the following characterization of base 2 notation multiplication as a guide. First, multiplication by a single digit is taken care of by: $x * 0 = 0$ and $x * 1 = x$. Next, if we know what $x * y$ is, it is enough to say what $x * y0$ and $x * y1$ are. The first is easy: $x * y0$ is $x * y$ with an extra 0 adjoined on the right. The second is only a little more complicated. $x * y1$ is the result of adding $x * y0$ and x.

2. $\mathbf{I}_2$ assigns the concatenation relation to CON, the set of odd-lengthed strings of 1's to EVEN, and the empty set to all other identifiers.
3. $\mathbf{I}_3$ assigns the concatenation relation to CON, the set of *all* strings of 1's to EVEN and to all other identifiers.
4. $\mathbf{I}_4$ assigns the empty relation to CON, the set of odd-lengthed strings of 1's to EVEN, and the empty set to all other identifiers.

Definition. If $\mathbf{I}$ is an interpretation and IDENTIFIER$(t_1, \ldots, t_j)$ is an atomic statement without variables, we call IDENTIFIER$(t_1, \ldots, t_j)$ *true under* $\mathbf{I}$ if $\langle t_1, \ldots, t_j \rangle \in \text{IDENTIFIER}^{\mathbf{I}}$.

If $X_1 \to X_2 \to \ldots \to X_p \to T$ is a statement without variables (with each X_i and T atomic), we call it *true under interpretation* $\mathbf{I}$ if at least one of the X_i is not true under $\mathbf{I}$, or else T is true under $\mathbf{I}$.

We call a statement with variables *true under interpretation* $\mathbf{I}$ if every substitution instance of it is true under $\mathbf{I}$.

For example, consider the two procedure statements

EVEN(Δ)
EVEN(x)→CON(x, 11, y)→EVEN(y).

You should verify that both are true under interpretation $\mathbf{I}_1$ above. Further, the second, but not the first statement is true under $\mathbf{I}_2$. Both are true under $\mathbf{I}_3$. And the second but not the first is true under $\mathbf{I}_4$.

Definition. Suppose $\mathbf{W}$ is a work space. We say an interpretation $\mathbf{I}$ is *in the work space* $\mathbf{W}$ if $\mathbf{I}$ assigns relations on the domain of $\mathbf{W}$ to identifiers, and $\mathbf{I}$ assigns to each reserved identifier of $\mathbf{W}$ the relation that $\mathbf{W}$ says the identifier represents.

For example, $\mathbf{I}_1$, $\mathbf{I}_2$, and $\mathbf{I}_3$ are interpretations in the basic work space of EFS(**str**({1})), but $\mathbf{I}_4$ is not.

Definition. Suppose we have a procedure P in work space $\mathbf{W}$. We say interpretation $\mathbf{I}$ is a *model* for P in $\mathbf{W}$ provided $\mathbf{I}$ is an interpretation in $\mathbf{W}$ under which every statement of P is true.

For example, $\mathbf{I}_1$ and $\mathbf{I}_3$ are both models for the procedure EVEN in the basic work space of EFS(**str**({1})). Now, the whole idea of the correctness proof presented earlier can be summarized as follows.

Proposition 1.6.1. *Let NAME be a procedure in work space* $\mathbf{W}$*, and let* $\mathbf{I}$ *be a model for NAME in* $\mathbf{W}$*. If* $\langle t_1, \ldots, t_n \rangle$ *is in the output of NAME in* $\mathbf{W}$*, then* $\langle t_1, \ldots, t_n \rangle$ *is in the relation* $NAME^{\mathbf{I}}$.

According to this proposition, to show that the output of procedure R is actually in the relation $\mathbf{R}$, it is enough to find a model for R in which the interpretation of the identifier R is $\mathbf{R}$. (See what conclusions this gives for EVEN using the models $\mathbf{I}_1$ and $\mathbf{I}_3$.)

Exercise 1.6.1. Give a proof of Proposition 1.6.1. It will be enough to show that every line of a trace of NAME is true under $\mathbf{I}$.

The correctness proof we gave for EVEN really amounted to showing that the interpretation $\mathbf{I}_1$ is a model, and then applying Proposition 1.6.1. The induction argument we gave earlier now becomes a special case of the proof of Proposition 1.6.1.

Exercise 1.6.2. Expand the basic work space of EFS(**str**({1})) by adding EVEN as a reserved identifier representing the set of even-lengthed words. Use the procedure ODD from §1.5 and prove its output is the set of odd-lengthed words.

Exercise 1.6.3. Prove that the output of the procedure FIB in the basic work space of EFS(**str**({1})), given in Exercise 1.4.4, really is the set of words whose lengths are Fibonacci numbers.

1.7 DENOTATIONAL SEMANTICS

In the last section we introduced models. Models can be used to characterize procedure behavior without resorting to details of computations. They provide the basis for an EFS denotational semantics. But a point of importance: Models are not unique. That is, there may be several models for a given procedure in a given work space. For example, in the previous section two distinct models for EVEN were introduced. In order to have a somewhat richer set of examples, we now consider the following procedure, in the basic work space of EFS(**str**($\{a, b\}$)).

SAMPLE (1):
  SAMPLE(a);
  SAMPLE(x) $\rightarrow$ CON(x, b, y) $\rightarrow$ SAMPLE(y).

It is easy to see the output of this is the set $\{a, ab, abb, abbb, \ldots\}$.

In specifying interpretations now, only SAMPLE and CON are significant. We assume all other identifiers are interpreted by the empty set. Further, we always take the interpretation of CON to be the concatenation relation on $\{a, b\}^*$. Then to say what interpretation we have in mind, all we need to do is assign a set of words to SAMPLE.

The following interpretations are, in fact, models for SAMPLE in the basic work space. Make sure you see why.

interpretation . . .	set assigned to SAMPLE
$\mathbf{I}_0$	$\{a, ab, abb, \ldots\}$
$\mathbf{I}_1$	$\{a, ab, abb, \ldots, ba, bab, babb, \ldots\}$
$\mathbf{I}_2$	$\{a, ab, abb, \ldots, ba, bab, babb, \ldots, bba, bbab, bbabb, \ldots\}$
$\mathbf{I}_3$	$\{a, ab, abb, \ldots, ba, bab, babb, \ldots, bbba, bbbab, bbbabb, \ldots\}$
$\mathbf{I}_4$	set of all words over $\{a, b\}$

For two models $\mathbf{I}$ and $\mathbf{J}$ of SAMPLE, let us write $\mathbf{I} \leq \mathbf{J}$ if $\text{SAMPLE}^{\mathbf{I}} \subseteq \text{SAMPLE}^{\mathbf{J}}$. Then $\mathbf{I}_0 \leq \mathbf{I}_1 \leq \mathbf{I}_4$, $\mathbf{I}_0 \leq \mathbf{I}_2 \leq \mathbf{I}_4$, $\mathbf{I}_0 \leq \mathbf{I}_3 \leq \mathbf{I}_4$, $\mathbf{I}_1 \leq \mathbf{I}_2$, $\mathbf{I}_1 \leq \mathbf{I}_3$, but neither of $\mathbf{I}_2 \leq \mathbf{I}_3$ nor $\mathbf{I}_3 \leq \mathbf{I}_2$ holds. We can represent this by the diagram

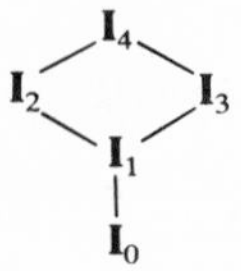

Since $\mathbf{I}_0$ assigns to SAMPLE the actual output of the procedure SAMPLE, Proposition 1.6.1 says $\mathbf{I}_0 \leq \mathbf{I}$ for any model of SAMPLE. In the $\leq$ ordering, $\mathbf{I}_0$ is the *smallest* model.

Exercise 1.7.1. Define other models of SAMPLE, and incorporate them into the diagram of models.

Exercise 1.7.2. Produce several models for EVEN, and a diagram showing their ordering.

As a matter of fact, whatever the procedure and whatever the work space, we can always find at least one model. The idea is to take as much to be true as possible.

Definition. Let $\mathbf{W}$ be a work space, and let P be a procedure. Consider the following interpretation. Interpret every reserved identifier as $\mathbf{W}$ specifies (thus we have an interpretation in $\mathbf{W}$). And, otherwise, if IDENTIFIER_ONE is used in P to represent a 1-place relation, interpret it by the universal set, consisting of all members of the domain. If IDENTIFIER_TWO is used in P to represent a 2-place relation, interpret it by the universal 2-place relation, the set of all pairs of members of the domain. And so on. (Identifiers not used in P can be interpreted arbitrarily.) We call such an interpretation a *trivial model* for P.

Calling the interpretation just defined a *model* requires justification, which is left for the next exercise. As an example, the interpretation $\mathbf{I}_4$ is a trivial model for SAMPLE.

Exercise 1.7.3. Suppose NAME is a procedure in work space **W** and let **I** be the trivial model defined for NAME above. Show that **I** actually is a model for NAME in **W**.

Thus, there is always at least one model for a procedure, and there may be several, some of which are not what we had in mind at all. What we show now is that there is always one "best" model. First, some preliminary work.

Definition. Suppose we have a family $\{\mathbf{I}_1, \mathbf{I}_2, \mathbf{I}_3, \ldots\}$ of interpretations (all involving the same domain). By the *intersection* of this family we mean the interpretation **J** that assigns to each identifier the intersection of the relations that $\mathbf{I}_1, \mathbf{I}_2, \mathbf{I}_3, \ldots$ assign to it. That is, $\langle t_1, \ldots, t_n \rangle$ will be in the relation IDENTIFIER$^{\mathbf{J}}$ if and only if $\langle t_1, \ldots, t_n \rangle$ is in the relation IDENTIFIER$^{\mathbf{I}_1} \cap$ IDENTIFIER$^{\mathbf{I}_2} \cap$ IDENTIFIER$^{\mathbf{I}_3} \cap \ldots$. We denote this interpretation by $\cap \{\mathbf{I}_1, \mathbf{I}_2, \mathbf{I}_3, \ldots\}$.

For example, consider the family $\{\mathbf{I}_0, \mathbf{I}_1, \mathbf{I}_2, \mathbf{I}_3, \mathbf{I}_4\}$ of models for SAMPLE given above. The intersection interpretation for this family happens to be $\mathbf{I}_0$ itself. Likewise, $\cap\{\mathbf{I}_2, \mathbf{I}_3\}$ is $\mathbf{I}_1$.

For reasons coming from elementary set theory, which we won't go into here, the intersection of an *empty* family of relations is a problematic case. However, in this section no difficulty will arise because every procedure has at least one model, the trivial one.

Exercise 1.7.4. Let $\{\mathbf{I}_1, \mathbf{I}_2, \ldots\}$ be a nonempty family of interpretations in work space **W**.

(a) Suppose that IDENTIFIER$(t_1, \ldots, t_n)$ is an atomic statement without variables. Show that IDENTIFIER$(t_1, \ldots, t_n)$ is true under $\cap\{\mathbf{I}_1, \mathbf{I}_2, \ldots\}$ if and only if it is true under each of $\mathbf{I}_1, \mathbf{I}_2, \ldots$.

(b) Show that $\cap\{\mathbf{I}_1, \mathbf{I}_2, \ldots\}$ is also an interpretation in **W**.

Now, the main fact we need is this.

Proposition 1.7.1. *Let NAME be a procedure in work space **W**, and let $\{\mathbf{I}_1, \mathbf{I}_2, \ldots\}$ be a nonempty family of models for NAME in **W**. Then $\cap\{\mathbf{I}_1, \mathbf{I}_2, \ldots\}$ is also a model for NAME in **W**.*

Proof. We only check the most significant item. Suppose that $X_1 \to X_2 \to \ldots \to X_p \to T$ is a substitution instance of a procedure statement of NAME. We show it is true in the intersection interpretation (here $X_1, X_2, \ldots, X_p, T$ are assumed to be atomic).

If any of $X_1, X_2, \ldots, X_p$ are not true in $\cap\{\mathbf{I}_1, \mathbf{I}_2 \ldots\}$ then $X_1 \to X_2 \to \ldots \to X_p \to T$ is automatically true, and we are done. So now suppose all of $X_1, X_2, \ldots, X_p$ are true in $\cap\{\mathbf{I}_1, \mathbf{I}_2 \ldots\}$. Then by Exercise 1.7.4, part (a), all of

$X_1, X_2, \ldots, X_p$ are true in each of $\mathbf{I}_1, \mathbf{I}_2, \ldots$. But $\{\mathbf{I}_1, \mathbf{I}_2 \ldots\}$ was a family of *models* for NAME, so $X_1 \to X_2 \to \ldots \to X_p \to T$ is true in each of $\mathbf{I}_1, \mathbf{I}_2, \ldots$. It follows that T is true in each of $\mathbf{I}_1, \mathbf{I}_2, \ldots$, so by Exercise 1.7.4, part (a) again, T is true in $\bigcap\{\mathbf{I}_1, \mathbf{I}_2, \ldots\}$. Thus, in any case, $X_1 \to X_2 \to \ldots \to X_p \to T$ is true in $\bigcap\{\mathbf{I}_1, \mathbf{I}_2 \ldots\}$.

Definition. By the *minimal model* for NAME in **W** we mean the intersection of all models for NAME in **W**.

Since every procedure has at least one model, the definition of minimal model is meaningful. By Proposition 1.7.1, the minimal model for a procedure NAME is actually a model for NAME. The minimal model for NAME in **W** is, of course, unique. And it is the one that "best" represents the behavior of the procedure NAME, in the following sense.

Proposition 1.7.2. *Let NAME be a procedure in work space* **W**, *and let* **M** *be the minimal model for NAME in* **W**. *The output of the procedure NAME is exactly the interpretation of NAME in the model* **M**, *NAME*$^{\mathbf{M}}$.

Proof. If $\langle t_1, \ldots, t_n \rangle$ is in the output of NAME in **W**, then $\langle t_1, \ldots, t_n \rangle$ is in NAME$^{\mathbf{M}}$ by Proposition 1.6.1 (and Proposition 1.7.1). Now suppose $\langle t_1, \ldots, t_n \rangle$ is *not* in the output of NAME in **W**. We must show that $\langle t_1, \ldots, t_n \rangle$ is not in NAME$^{\mathbf{M}}$. To do this, it is enough to show there is at least one model **I** for NAME in **W** with $\langle t_1, \ldots, t_n \rangle$ not in NAME$^{\mathbf{I}}$.

Exercise 1.7.5. (Continuing the proof of Proposition 1.7.2)

(a) Define an interpretation **I** as follows. For each identifier, let $\langle t_1, \ldots, t_n \rangle$ be in the relation IDENTIFIER$^{\mathbf{I}}$ just in case there is some trace of NAME whose last line is IDENTIFIER$(t_1, \ldots, t_n)$. Show **I** is a model for NAME in **W**.
(b) Finish the proof of Proposition 1.7.2.
(c) Show the model **I** constructed in part (a) *is* the minimal model.

1.8 BACKGROUND

We end each chapter with a Background section, like this one, providing references to the literature where key concepts discussed in the chapter originated. This does not mean these references always contain the "best" presentation of the material. Often, later work considerably clarified matters. But we should know where ideas came from, and when, as well as where they are now.

The EFS language presented in this chapter derives directly from [Smullyan 1956] and especially from [Smullyan 1961], where it was used to give an elegant and simple definition of the class of recursive functions. See §5.15, Background, in Chapter 5 for more on recursive functions. Independently, and some years later, the idea that logic-based languages could be of use in programming gained currency; see [Kowalski 1974]. So-called *logic programming,* in its simplest form, is essentially the same as Smullyan's system except for the choice of data structure. Smullyan used the structure of *words* presented in this chapter; logic programming takes as its domain the collection of *terms* of formal logic, getting a data structure we do not consider here, though it is sketched briefly in Chapter 2, §2.10. [Lloyd 1984] summarizes the theory of logic programming very nicely.

The "real" programming language that comes closest to an EFS-style language is PROLOG, based on ideas from [Kowalski 1974] and described in [Roussel 1975]. Implementations are now available for both microcomputers and for mainframes. There are minor syntax differences between PROLOG and EFS. What we write as $A \rightarrow B \rightarrow C \rightarrow D$ in EFS would be written as $D \leftarrow (A \,\&\, B \,\&\, C)$, or some variant, in PROLOG. The data structure assumed in PROLOG is not one considered here. But most importantly, PROLOG is given a deterministic operational semantics, which leads to issues of control that are not relevant in EFS. Nonetheless, the underlying ideas of PROLOG will be very familiar to a reader of this book. If a PROLOG implementation is available to the reader, for goodness sake, make use of it. There are a number of fairly specific books intended for PROLOG programmers, [Clocksin and Mellish 1981] is something of a standard. A more general work that we strongly recommend is [Kowalski 1979].

Minimal models were first defined, very briefly, in [Smullyan 1956A]. They were originally intended to provide yet another characterization of the class of recursive functions. The minimal model construction was rediscovered in the logic programming context, and its importance was recognized as providing a denotational semantics, in [van Emden and Kowalski 1976]. It was further investigated in [Apt and van Emden 1982].

REFERENCES

[Apt and van Emden 1982] Contributions to the theory of logic programming, K. Apt and M. van Emden, *Journal of the Association for Computing Machinery,* Vol. 29, pp. 841–862, 1982.

[Clocksin and Mellish 1981] *Programming in PROLOG,* W. F. Clocksin and C. S. Mellish, Springer-Verlag, Berlin, 1981.

[Kowalski 1974] Predicate logic as a programming language, R. Kowalski, *Proc. IFIP Congress 1974,* North Holland Publishing Co., Amsterdam, pp. 569–574, 1974.

[Kowalski 1979] *Logic for Problem Solving,* R. Kowalski, Elsevier-North Holland Publishing Co., New York, 1979.

[Lloyd 1984] *Foundations of Logic Programming,* J. Lloyd, Springer-Verlag, Berlin, 1984.

[Roussel 1975] *PROLOG: Manuel de Reference et d'Utilisation,* P. Roussel, Groupe d'Intelligence Artificielle, U.E.R., de Luminy, Université d'Aix-Marseille, Sept. 1975.

[Smullyan 1956] Elementary formal systems (abstract), R. Smullyan, *Bulletin of the American Mathematical Society,* Vol. 62, p. 600, 1956.

[Smullyan 1956A] On definability by recursion (abstract), R. Smullyan, *Bulletin of the American Mathematical Society,* Vol. 62, p. 601, 1956.

[Smullyan 1961] *Theory of Formal Systems,* revised edition, R. Smullyan, Princeton University Press, Princeton, N.J., 1961.

[van Emden and Kowalski 1976] The Semantics of predicate logic as a programming language, M. Van Emden and R. Kowalski, *Journal of the Association for Computing Machinery,* Vol. 23, pp. 733–742, 1976.

2

THE FAMILY OF EFS LANGUAGES

2.1 EFS LANGUAGES

Definition. By a *data structure* we mean a $k+1$ tuple $\langle \boldsymbol{D}; \boldsymbol{R}_1, \ldots, \boldsymbol{R}_k \rangle$ where $\mathbf{D}$ is a collection called the *domain* and $\mathbf{R}_1, \ldots, \mathbf{R}_k$ are relations on $\mathbf{D}$.

For example, in Chapter 1, $\mathbf{str}(L) = \langle L^*; \mathbf{CON}_L \rangle$ was the data structure considered. Now, for every choice of data structure, we want to set up an EFS language that uses it. In this section we give the necessary definitions, and we devote the next several sections to examples.

To be precise, what we are calling a data structure is really an *algebraic* data structure. We don't say how the domain and the relations are created, we simply assume we have them and that they behave in ways that can be properly formulated and used in proofs about program behavior. In practice we will not be formal about such things but will assume an intuitive understanding of the properties of numbers, trees, and so on, just as we did with words in Chapter 1. In fact, almost all our examples will be of *recursive* data structures; the domain can be *generated* from a finite set of initial objects using a finite number of operations that are embodied in the given relations. For example, words over L are generated from the empty word using the operation of adding a single letter, a simple form of concatenation. Recursive data structures in general have a well-developed theory, which we do not go into here because it would take us too far afield. So, although our examples are usually of recursive data structures, we make no direct use of that fact.

In Chapter 1, procedure statements could contain constants, which were character strings over an alphabet L. In this chapter, the EFS language we set up for the data structure $\langle \mathbf{D}; \mathbf{R}_1, \ldots, \mathbf{R}_k \rangle$ will allow *members of* $\boldsymbol{D}$ to be used as constants in procedure statements. For example, when working with a data structure whose domain is numbers, numbers may appear as constants in procedure statements. Of course, this is all highly mythological; how can abstract objects like numbers be used in such a concrete way? What we really should say is this. An EFS language that is supposed to work with numbers should be thought of as being implemented on an idealized computer. As part of that implementation, some internal representation for numbers has been chosen. What actually appears as a constant in a procedure statement is some configuration that we and the computer can understand as designating which number representative the computer will use when working with that procedure statement. But it will complicate our discussions if we have to keep saying all this each time a procedure statement is discussed. It is simpler for us to talk about numbers themselves, and let you mentally expand that, if you wish.

The next problem is with the given relations of a data structure. We would like them to be, at least in principle, subject to implementation on computers as we know them. This may not always be the case, but we must postpone the issue for now. We simply *assume* our language will be implemented on a computer capable of generating the necessary relations somehow, and we go on from this point. In Chapter 4 we discuss criteria for distinguishing implementable from unimplementable data structures. For now, we allow anything.

Now we present a generalization of EFS language from Chapter 1. Let **S** be the data structure $\langle \mathbf{D}; \mathbf{R}_1, \ldots, \mathbf{R}_k \rangle$. We define the language EFS(**S**).

Identifiers, Variables, and Punctuation. The same as in Chapter 1.

Terms. The terms of EFS(**S**) are the variables and members of **D**.

Statements, Acceptable Statements. Notions of atomic statement and statement are exactly as in Chapter 1. We still take some identifiers to be reserved, and we still call a statement acceptable if no reserved identifier is used in the assignment position.

Work Space. By a work space **W** we mean (1) a specification of a domain **D** (not necessarily consisting of words now), (2) a list of reserved identifiers, and (3) a specification of what relations on the domain **D** the reserved identifiers represent (called the *given* relations of the work space).

The **basic work space** of EFS(**S**), where $\mathbf{S} = \langle \mathbf{D}; \mathbf{R}_1, \ldots, \mathbf{R}_k \rangle$ has domain **D**, given relations $\mathbf{R}_1, \ldots, \mathbf{R}_k$, and reserved identifiers that will be specified on a case by case basis as we give our examples in the following sections.

Procedures. The same definition as in Chapter 1, §1.3.

Computation, Output, Trace. The same definition as in Chapter 1, §1.4.

We continue the practice of expanding a work space by writing a procedure whose output is to be a given relation for future work.

In the next few sections we pick certain data structures of particular interest, and we introduce "standard" identifiers for the given relations, much as we did with CON in Chapter 1. This will give us a collection of basic work spaces. Most work spaces we look at are either these, or expansions of them.

2.2 NUMBERS

Definition. **N** is the set of nonnegative integers, $\{0, 1, 2, \ldots\}$. The *successor relation* **SUC** on **N** is the set of ordered pairs of the form $\langle a, b \rangle$ where b is the integer immediately following a. $\langle \mathbf{N}; \mathbf{SUC} \rangle$ is a data structure, which we call **integer**. The language EFS(**integer**) is as defined in the last section. By the *basic work space* of EFS(**integer**) we mean the one with: domain **N**; the identifier SUC reserved; SUC representing **SUC**.

For example, $\langle 0,1 \rangle$, $\langle 1,2 \rangle$, $\langle 2,3 \rangle, \ldots$ are in the successor relation **SUC**. We give examples of procedures in EFS(**integer**) below. One point first, however. The definition in §2.1 says we can use members of **N** themselves as terms in procedure statements. Of course, our procedure statements actually must appear in print in this book, so we need some way of indicating what numbers we are considering. For this purpose we use conventional base 10 notation. Please understand, though, that our procedures should not be thought of as manipulating base 10 *notation*. **SUC** is a relation on *numbers*.

The following procedure is written in the basic work space of EFS(**integer**). It is intended to have the addition relation as output.

SUM (3):
  SUM(x, 0, x);
  SUM(x, y, z)$\rightarrow$SUC(y, u)$\rightarrow$SUC(z, v)$\rightarrow$SUM(x, u, v).

Here is a trace of SUM showing that $\langle 3,1,4 \rangle$ is in the output, which is reasonable since $3+1=4$.

SUM(3, 0, 3)
SUC(0, 1)
SUC(3, 4)
SUM(3, 0, 3)$\rightarrow$SUC(0, 1)$\rightarrow$SUC(3, 4)$\rightarrow$SUM(3, 1, 4)
SUM(3, 1, 4)

We do not attempt to prove here that the output is exactly the addition relation. We leave that to later in the chapter. It is true, however, and we take it for granted in what follows. We make similar assumptions about the various procedures below. We now present a sequence of procedures that will be of use to us. It is understood that, every time we write a procedure, the work space is extended using it. Thus, each procedure output is available when writing later procedures. In particular, SUM is available now.

The equality relation, which is rather trivial.

EQUAL (2):
  EQUAL(x, x).

The not-equal-to-zero property.

NOT_ZERO (1):
  SUC(x, y)$\rightarrow$NOT_ZERO(y).

The less-than relation on integers.

LESS (2):
  SUM(x, y, z)→NOT_ZERO(y)→LESS(x, z).

The relation of being distinct integers.

NOT_EQUAL (2):
  LESS(x, y)→NOT_EQUAL(x, y);
  LESS(y, x)→NOT_EQUAL(x, y).

The less-than-or-equals relation.

LESS_OR_EQUAL (2):
  LESS(x, y)→LESS_OR_EQUAL(x, y);
  EQUAL(x, y)→LESS_OR_EQUAL(x, y).

The multiplication relation on integers, that is, the set of triples $\langle a, b, c\rangle$ where c is the product of a and b.

PROD (3):
  PROD(x, 0, 0);
  PROD(x, y, z)→SUC(y, u)→SUM(x, z, v)→PROD(x, u, v).

The relation that holds between a, b, q, and r when q is the quotient and r is the remainder on dividing a by nonzero b.

QUOTIENT_AND_REMAINDER (4):
  NOT_ZERO(b)→PROD(q, b, x)→SUM(x, r, a)
    →LESS(r, b)→QUOTIENT_AND_REMAINDER(a, b, q, r).

From this we easily get two related procedures.

QUOTIENT (3):
  QUOTIENT_AND_REMAINDER(a, b, q, r)→QUOTIENT(a, b, q).

REMAINDER (3):
  QUOTIENT_AND_REMAINDER(a, b, q, r)→REMAINDER(a, b, r).

The relation of exact division, b divides a.

DIVIDES (2):
  REMAINDER(a, b, 0)→DIVIDES(b, a).

This list should be sufficient to give a feeling for procedures in EFS(**integer**). The following exercises ask you to extend the list by writing additional procedures.

Exercise 2.2.1. Write procedures for the following (making use of those above if necessary).

(a) NOT_DIVIDES for the relation: y does not exactly divide x.

(b) EVEN for the set of even numbers.
(c) NOT_EVEN for the set of numbers that are not even.
(d) POWER_OF_TWO for the set of powers of two.
(e) NOT_POWER_OF_TWO for the set of numbers that are not powers of two. (*Hint*: A number is not a power of two if and only if it is strictly between a power of two and the next power of two.)
(f) POWER for the relation to-the-power-of.
(g) NOT_PRIME for the set of nonprimes. (We will consider the set of primes itself in §2.7, but you might think about it now.)
(h) RELPRIME for the relation: a and b are relatively prime. (*Hint*: A result of number theory says that a and b are relatively prime just when there are signed integers x and y such that $ax + by = 1$.)
(i) BASE_TWO_LENGTH for the relation: n is the number of digits in the base 2 expression for x.

2.3 BINARY TREES

Definition. Let **A** be some collection of things, which we call *atoms,* and which are not themselves trees. **T(A)** is the collection of finite, ordered, binary *trees* with atoms as leaves.

Each node can have two immediate successors (binary), a left and a right, and we can tell left from right (ordered). For example, if *a, b, c, d,* and *e* are members of **A**, that is, atoms, then the following are members of **T(A)**:

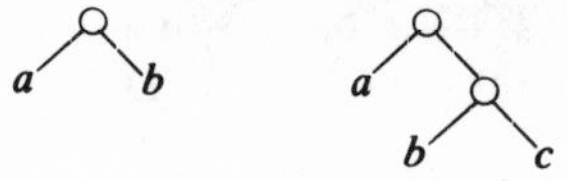

and

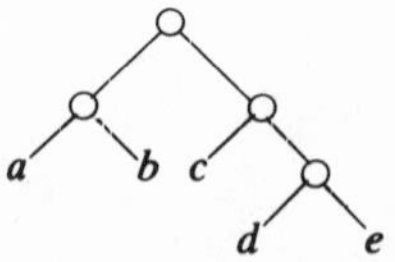

We also think of atoms themselves as honorary trees, identifying atom a with the trivial tree whose only leaf is a.

If we have a nontrivial tree, we can talk about its left and right immediate subtrees. The following indicates what we mean by this.

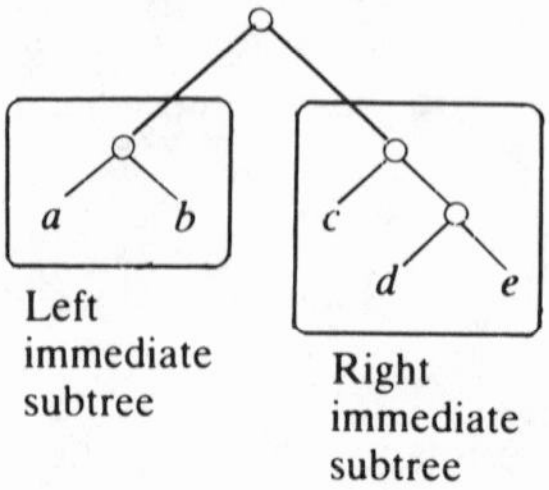

The left and right immediate subtrees of

are the trivial trees *a* and *b* which, in turn, have no left or right immediate subtrees.

Definition. By **tree(A)** we mean the data structure ⟨**T(A)**; **IMSUB**, **ATOM**⟩ where the domain is **T(A)**; **IMSUB** is the relation consisting of all triples ⟨*x*, *y*, *z*⟩ where *x* is a tree and *y* and *z* are its immediate left and right subtrees; and **ATOM** is the property of being an atom (i.e., **ATOM** = **A**).

By the *basic work space* of EFS(**tree(A)**) we mean the one with **T(A)** as domain, IMSUB and ATOM as reserved identifiers and with IMSUB representing **IMSUB** and ATOM representing **ATOM**.

Before we give an example of a procedure, there is a point to be considered. Members of **T(A)** are constants of EFS(**tree(A)**); we need some way of representing them on a printed page. A similar issue came up in §2.2 with numbers. Well, we will continue representing them in the schematic way we did above. When a procedure statement is displayed in this section, it may contain a two-dimensional representation for a tree, indicating in print what tree is being used at this point. Do not confuse the printed representation for a tree with the tree itself, or with a possible computer representation. The given relations are relations on trees.

Here is an example of a procedure in the basic work space of EFS(**tree(A)**). The output is intended to be the relation: all ordered pairs ⟨*x*, *y*⟩ where *y* is a tree and *x* is an atom occurring in it.

ATOM_OF (2):
  ATOM(*x*) → ATOM_OF(*x*, *x*):
  ATOM_OF(*x*, *y*) → IMSUB(*w*, *y*, *z*) → ATOM_OF(*x*, *w*);
  ATOM_OF(*x*, *z*) → IMSUB(*w*, *y*, *z*) → ATOM_OF(*x*, *w*).

We give a trace of this procedure, under the assumption that the set of atoms, **A**, is the set of lower case letters, *a*, *b*, ... We show one of the outputs is ⟨*a*, 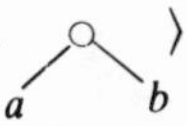 ⟩

ATOM(*a*)
ATOM(*a*) → ATOM_OF(*a*, *a*)
ATOM_OF(*a*, *a*)
IMSUB( , *a*, *b*)
ATOM_OF(*a*, *a*) → IMSUB(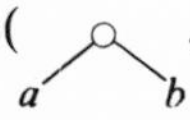 , *a*, *b*) → ATOM_OF(*a*, 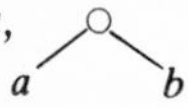)
ATOM_OF(*a*, 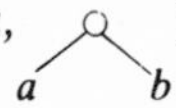)

You might try extending this to show $\langle a,$ 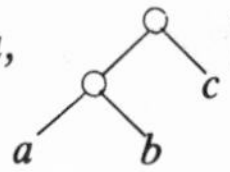

$\rangle$ is also an output.

Definition. We say tree T_1 is a *subtree* of T_2 if $T_1 = T_2$ or T_1 is an immediate subtree (left or right) of an immediate subtree of . . . of T_2.

For example, 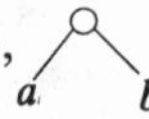

is a subtree of 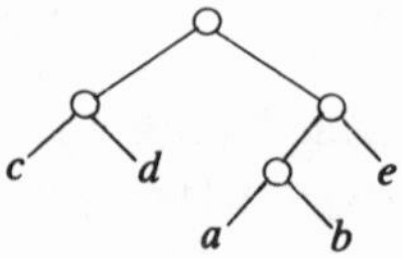

Exercise 2.3.1. Write a procedure SUBTREE for the relation: is a subtree of.

Definition. Number representatives can be introduced through the following simple device. We choose two atoms (assume there are at least two), say a and b, to be used for this purpose, and we say

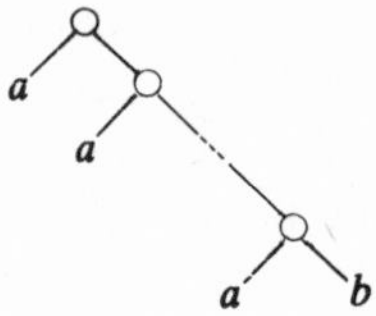

stands for the number n, where there are n occurrences of a. The trivial tree with the single leaf b stands for 0. We call trees of this form *number representatives.*

Exercise 2.3.2. Write a procedure NUM whose output is the set of number representatives.

Exercise 2.3.3. Write a procedure SUC for the "successor" relation on number representatives.

Exercise 2.3.4. Write a procedure LESS for the "less-than" relation on number representatives. Also LESS_OR_EQUAL for the "less-than-or-equals" relation.

Definition. By the *depth* of a tree we mean the length of the longest branch (starting our count with 0, that is, a trivial tree has depth 0).

Exercise 2.3.5. Write a procedure DEPTH_OF whose output is the set of ordered pairs $\langle x, y \rangle$ where x is a tree and y is a number representative whose "value" is the depth of x.

Exercise 2.3.6. Write a procedure SAME_DEPTH whose output is the relation: x and y are trees of the same depth.

Similarly write a procedure NOT_SAME_DEPTH whose output is the relation: x and y are trees of different depths.

Of course, two trees can be of the same depth without being the same tree. The relation of inequality is not programmable because we have no mechanism for distinguishing between atoms in the language. For this we must assume a richer basic work space.

Definition. By **tree(A, $\neq$)** we mean the structure $\langle$**T(A)**; **IMSUB**, **ATOM**, **NOTEQ**$\rangle$ where **T(A)**, **IMSUB**, and **ATOM** are as above, and **NOTEQ** is the not-equals relation on atoms. The basic work space of EFS(**tree(A, $\neq$)**) is like that of **tree(A)**, but with NOTEQ as an added reserved identifier, representing **NOTEQ**.

Exercise 2.3.7. Write a procedure NOT_EQUAL in the basic work space of EFS(**tree(A, $\neq$)**) whose output is the relation holding between distinct trees.

Exercise 2.3.8. Write a procedure REPLACE whose output is the set of 4-tuples $\langle x, y, z, w\rangle$ where x is a tree, y is an atom, z is a tree, and w is the result of replacing all occurrences of atom y in tree x by occurrences of tree z.

2.4 SETS

Definition. Again let **A** be some collection of things called atoms, but this time we assume members of **A** are not sets. Let **S(A)** be the collection consisting of members of **A**, finite sets of members of **A**, finite sets made up of members of **A**, and finite sets of members of **A**, etc. In short, members of **S(A)** are all the things we can get by starting with atoms and repeating the operation, finite set of.

For example, if a, b, and c are atoms, the following are all members of **S(A)**:

a, b, c
$\{a, b\}$
$\{c, \{a, b, c\}\}$
$\{\{b\}, \{c, \{a, b\}\}\}$.

Sets are familiar objects to mathematicians, but they are not so common as data objects in computer science. If you wish, you can think of a set as a list of its members, in which order and repetition don't matter. Thinking of it this way would be backwards to a mathematician, but a computer representation of a set is often something like this. One of the reasons sets are relatively uncommon as data objects is that, in practice, ignoring order and repetition can take a great deal of work.

Definition. By **set(A)** we mean the data structure $\langle$**S(A)**; **ADDMEM**, **ATOM**$\rangle$ where **S(A)** is as above, **ADDMEM** is the set of triples $\langle x, y, z\rangle$ where x is a set and z is the result of adding y as a member to x, and **ATOM** = **A** is the property of being an atom. The language suitable for this data structure is EFS(**set**(**A**)).

By the *basic work space* of EFS(**set**(**A**)) we mean the one with **S**(**A**) as domain, ADDMEM and ATOM as reserved identifiers, representing **ADDMEM** and **ATOM** respectively.

For example $\langle \{a, b\}, c, \{a, b, c\} \rangle$ is in the relation **ADDMEM**. Note: If y is already in x, adding y to x leaves x unchanged. We are talking here about sets, not about multisets where members occur with a multiplicity factor. Once again we have the problem of representing constants, sets, on the printed page. We will continue the "standard" scheme used above, writing $\{x, y, z\}$ to denote the set whose members are $x, y,$ and z. This notation is not unique; $\{y, z, x\}$ and $\{x, x, y, z\}$ also denote this set.

The first example of a procedure we give parallels the first one in §2.3. The output is intended to be the relation: all pairs $\langle x, y \rangle$ where y is a set (or atom) and x is an atom occurring in it, or in a member of it, or in a member of a member of it, etc.

ATOM_OF (2):
  ATOM(x)→ATOM_OF(x, x);
  ATOM_OF(x, y)→ADDMEM(y, z, w)→ATOM_OF(x, w):
  ATOM_OF(x, z)→ADDMEM(y, z, w)→ATOM_OF(x, w).

This time we leave it to you to run some traces of the procedure and get a feeling for it.

Exercise 2.4.1. Write a procedure MEMBER_OF for the relation: is a member of. *Note*: If c is an atom, it has no members, but c is a member of $\{c\}$.

Exercise 2.4.2. Write a procedure SET for the property: is a set; that is, is not an atom.

Exercise 2.4.3. Write a procedure SUBSET for the relation: set x is a subset of set y.

Exercise 2.4.4. Write a procedure UNION for the relation: set x is the union of sets y and z.

Exercise 2.4.5. Write a procedure ADD_TO_MEMBERS_OF for the relation: x is a set of sets and z is the set which results when y is added to each member of x.

Exercise 2.4.6. Write a procedure POWER for the relation: Set x is the collection of all subsets of set y.

Once again, number representatives can be introduced. There are many ways of doing this; we use a system of von Neumann in which the representative of n is taken to be the set of representatives of all smaller numbers. Let us write $\bar{n}$ for the set representing the number n. Since 0 is the smallest (nonnegative) number, its representative has no members, that is, it is the empty set $\{\ \}$. Thus

$$\bar{0} = \{\ \} = \emptyset.$$

After this,

$$\bar{1} = \{\bar{0}\} = \{\{\ \}\}$$
$$\bar{2} = \{\bar{0}, \bar{1}\} = \{\{\ \}, \{\{\ \}\}\}$$

etc. In general,

$$\begin{aligned} \overline{n+1} &= \{\bar{0}, \bar{1}, \ldots, \overline{n-1}, \bar{n}\} \\ &= \{\bar{0}, \bar{1}, \ldots, \overline{n-1}\} \cup \{\bar{n}\} \\ &= \bar{n} \cup \{\bar{n}\}. \end{aligned}$$

Definition. The representative of 0 is $\emptyset$. If $\bar{n}$ is the representative of n, the representative of $n+1$ is $\bar{n} \cup \{\bar{n}\}$.

Exercise 2.4.7. Write a procedure NUM whose output is the set of number representatives.

Exercise 2.4.8. Write a procedure SUC for the "successor" relation on number representatives.

Because of the way number representatives were chosen, the subset relation on them is the less-than-or-equals relation, the member-of relation is the less-than relation, the union operation is the maximum-of operation, and the intersection operation is the minimum-of operation.

Definition. By the *depth* of a set or atom s we mean the largest number of times the member-of operation can be applied, starting with s.

For example, atoms have depth 0. For a more complicated example, suppose a, b, and c are atoms, and consider the set $\{a, \{b\}, \{b, \{c\}\}\}$. We have the following tree, where at each level we display the members of the parents.

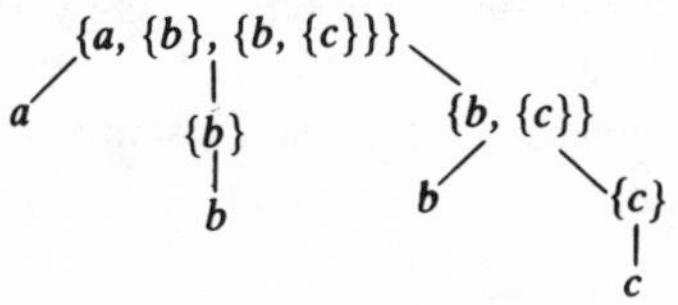

The longest branch here is on the right: It descends three levels to the end. Then the depth of $\{a, \{b\}, \{b, \{c\}\}\}$ is 3.

Exercise 2.4.9. Write a procedure DEPTH_OF whose output is the set of ordered pairs $\langle x, y \rangle$ where y is a number representative whose "value" is the depth of x.

Exercise 2.4.10. Write a procedure SAME_DEPTH whose output is the relation: x and y are of the same depth. Also write a procedure NOT_SAME_DEPTH whose output is the relation: x and y are of different depths.

You wrote a MEMBER_OF procedure as Exercise 2.4.1. The complementary relation not-a-member-of cannot be programmed in the present language. If b and c are atoms, to conclude that b is not a member of $\{c\}$ we would need to know that b and c are different. But we were given no such information about atoms to work with.

Definition. By **set**(**A**, $\neq$) we mean the structure $\langle$**S**(**A**); **ADDMEM**, **ATOM**, **NOTEQ**$\rangle$ where **S**(**A**), **ADDMEM**, and **ATOM** are as above, and **NOTEQ** is the not-equals relation on atoms. The basic work space of EFS(**set**(**A**, $\neq$)) is like that of **set**(**A**), but with NOTEQ as an added reserved identifier, representing **NOTEQ**.

With this extra machinery a procedure can be written for not-a-member-of. Since it is somewhat tricky we present it, rather making it an exercise. The idea is this.

To say x and y are different is to say:

1. Both are atoms, but are distinct, or
2. One is an atom, but the other is not, or
3. Both are sets, but one has a member which the other lacks.

Items (1) and (2) pose no problem. The key feature of item (3) is that it reduces a question about x and y to questions about members, which are of lower depth, thus simplifying the problem.

Now to say x is not a member of y is to say:

1. y is an atom, or is empty, or
2. y is a nonempty set but x is different than any member.

Here item (1) is straightforward, whereas item (2) reduces a question about x and y to questions about x and members of y, again a reduction in depth. And we can check that x is different from the members of y by comparing it with the members, one at a time. Thus, having compared x with one member of y, we have fewer things left to compare it with, again a simplification of the problem.

With this as background, here is the procedure. We will prove the correctness of this procedure in §2.8, and its completeness in §2.9.

NOT_MEMBER_OF (2):

  ATOM(x)$\rightarrow$ATOM(y)$\rightarrow$NOTEQ(x, y)$\rightarrow$DIFFERENT(x, y);

  ATOM(x)$\rightarrow$SET(y)$\rightarrow$DIFFERENT(x, y);

SET(x)→ATOM(y)→DIFFERENT(x, y);
SET(x)→SET(y)→MEMBER_OF(z, x)
→NOT_MEMBER_OF(z, y)→DIFFERENT(x, y);
SET(x)→SET(y)→NOT_MEMBER_OF(z, x)
→MEMBER_OF(z, y)→DIFFERENT(x, y);
ATOM(y)→NOT_MEMBER_OF(x, y);
NOT_MEMBER_OF(x, { });
NOT_MEMBER_OF(x, y)→DIFFERENT(x, w)
→ADDMEM(y, w, z)→NOT_MEMBER_OF(x, z).

Exercise 2.4.11. Write a NOT_EQUAL procedure for the unequals relation on atoms and sets.

Exercise 2.4.12. Write a NOT_SUBSET procedure.

Exercise 2.4.13. Write a procedure NOT_NUM whose output is the set of things that are not number representatives. *Hint*: A number representative represents its own depth.

2.5 A RELATIONAL DATA BASE

Let **NAMES** be a collection of names of people. And suppose we have the following relations on **NAMES**: **MALE**, the property of being the name of a male; **FEMALE**, the property of being the name of a female; **PARENT_OF**, the relation is-a-parent-of, and **NOTEQ**, the relation of being names of different people. We consider the data structure ⟨**NAMES**; **MALE**, **FEMALE**, **PARENT_OF**, **NOTEQ**⟩. Unlike the ones of the last few sections, this structure will not come up again in the book, so we introduce no special name for it.

The basic work space of EFS(⟨**NAMES**; **MALE**, **FEMALE**, **PARENT_OF**, **NOTEQ**⟩) uses MALE, FEMALE, PARENT_OF, and NOTEQ as reserved identifiers for **MALE**, **FEMALE**, **PARENT_OF**, and **NOTEQ**, respectively.

Here is a procedure for the relation of being an ancestor.

ANCESTOR (2):
PARENT_OF(x, y)→ANCESTOR(x, y);
ANCESTOR(x, y)→PARENT_OF(y, z)→ANCESTOR(x, z).

What would be the effect of replacing the second procedure staetment with

ANCESTOR(x, y)→ANCESTOR(y, z)→ANCESTOR(x, z)?

Exercise 2.5.1. Write procedures for the following:

(a) SIBLING, the brother-or-sister-of relation
(b) BROTHER_OF and SISTER_OF
(c) AUNT_OF and UNCLE_OF
(d) COUSIN_OF

2.6 LOGICAL NOTATION

In the procedures of the last several sections, certain simple patterns of instructions came up repeatedly. It will be useful to introduce a special notation for these patterns. The notation comes from mathematical logic. The presentation in this section is not a formal one; we return to the material more formally in Chapter 5.

Definition. We call a relation that is the output of a procedure a *generated* relation. Of course, this is relative to a choice of EFS language.

Let **W** be a work space, fixed for the time being.

To begin with, we can "switch around the places" in generated relation. For example, suppose, R is a reserved identifier of **W**, representing a 4-place relation **R**. Suppose we define a relation **S** by $\langle x, y, z, w\rangle \in \mathbf{S}$ if $\langle w, y, z, x\rangle \in \mathbf{R}$. It is easy to show that **S** is also generated. Simply take an unreserved identifier, say S, and use the procedure

S (4):
$$R(w, y, z, x) \rightarrow S(x, y, z, w).$$

From now on we will "shortcut" the full writing of such a procedure by simply saying: let S be given by

$$S(x, y, z, w) \Leftrightarrow R(w, y, z, x).$$

Next, we can "add places" to a generated relation. This time suppose R is a reserved identifier representing a 2-place relation, and S is unreserved. We may write: let S be given by

$$S(x, y, z) \Leftrightarrow R(x, y).$$

We think of this as short for the procedure

S (3):
$$R(x, y) \rightarrow S(x, y, z).$$

Similarly we can "insert constants." Suppose R is a reserved identifier representing a 3-place relation, and S is unreserved. Let c be some constant, that is, some member of the work space domain. We write: Let S be given by

$$S(x, y) \Leftrightarrow R(x, c, y).$$

This is short for

S (2):
$$R(x, c, y) \rightarrow S(x, y).$$

Of course, various combinations of the items above may appear. It should be obvious how to deal with them.

Next, suppose that R and S are both reserved identifiers of **W**, both representing generated n-place relations, **R** and **S** respectively. Define n-place relations **T** and **V** as follows.

$\langle x_1, \ldots, x_n \rangle \in \mathbf{T}$ if $\langle x_1, \ldots, x_n \rangle \in \mathbf{R}$ and $\langle x_1, \ldots, x_n \rangle \in \mathbf{S}$

$\langle x_1, \ldots, x_n \rangle \in \mathbf{V}$ if $\langle x_1, \ldots, x_n \rangle \in \mathbf{R}$ or $\langle x_1, \ldots, x_n \rangle \in \mathbf{S}$, or both.

In set-theoretic terms, $\mathbf{T} = \mathbf{R} \cap \mathbf{S}$ and $\mathbf{V} = \mathbf{R} \cup \mathbf{S}$. Both **T** and **V** are also generated relations. Let T and V be unreserved identifiers. Here is a procedure whose output is **T**.

$T\ (n)$:
$$R(x_1, \ldots, x_n) \rightarrow S(x_1, \ldots, x_n) \rightarrow T(x_1, \ldots, x_n).$$

And here is a procedure whose output is **V**.

$V\ (n)$:
$$R(x_1, \ldots, x_n) \rightarrow V(x_1, \ldots, x_n);$$
$$S(x_1, \ldots, x_n) \rightarrow V(x_1, \ldots, x_n).$$

In formal logic, the symbols "$\wedge$" and "$\vee$" are often used for "and" and "or." From now on we will generally shortcut writing procedures like those above and simply say: let T and V be given by

$$T(x_1, \ldots, x_n) \Leftrightarrow R(x_1, \ldots, x_n) \wedge S(x_1, \ldots, x_n)$$
$$V(x_1, \ldots, x_n) \Leftrightarrow R(x_1, \ldots, x_n) \vee S(x_1, \ldots, x_n).$$

In fact, we will make use of this notation in a somewhat more general setting still. We will not insist that R and S always represent relations with the *same* number of places, since we can always "pad out" the shorter one by adding places as described earlier. For example, suppose that R represents a 2-place and S a 4-place relation, and we write

$$T(x, y, z, w) \Leftrightarrow R(x, w) \wedge S(x, y, z, w).$$

We can take this as short for

$T(4)$:
$$R(x, w) \rightarrow S(x, y, z, w) \rightarrow T(x, y, z, w).$$

Similar remarks apply to "or", of course.

Finally, suppose that R is a reserved identifier representing an $n+1$ place relation **R**. Define an n-place relation **S** by

$\langle x_1, \ldots, x_n \rangle \in \mathbf{S}$ if $\langle x_1, \ldots, x_n, y \rangle \in \mathbf{R}$ for some y.

The relation **S** is sometimes called a *projection* of **R**. The term *projection* is from geometry. The following diagram illustrates a projection of a 2-place relation $\mathbf{R}(x, y)$ and should account for the use of the term.

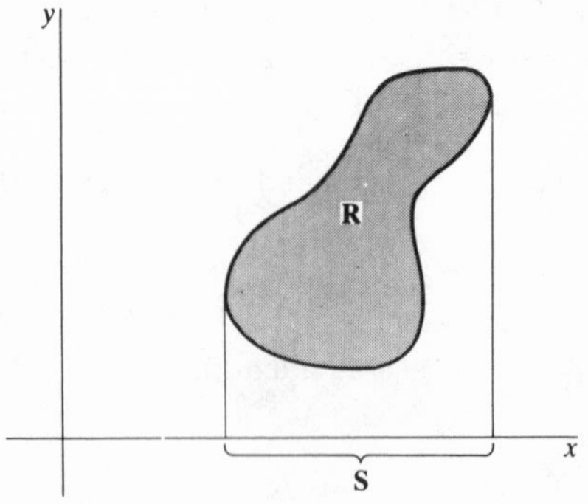

Figure 2.1

If **R** is a generated relation, **S** is also generated. Let S be unreserved, and use the procedure

$S\,(n)$:

$$R(x_1, \ldots, x_n, y) \rightarrow S(x_1, \ldots, x_n).$$

In formal logic, the *existential quantifier* "$\exists$" is generally used to stand for "there exists." Thus "$(\exists y)$" can be read "there exists a y such that" or "for some y." Then, we will shortcut the display of the procedure above by writing: let

$$S(x_1, \ldots, x_n) \Leftrightarrow (\exists y) R(x_1, \ldots, x_n, y).$$

The notation of logic introduced above will generally appear in fairly elaborate combinations. For example, in §2.2 we began with the basic work space of EFS(**integer**), and gradually extended it. Let us consider a still-further extension. Suppose we want to show the following 3-place relation is generated: Either x is a multiple of 3 or x is between y and z. Well, take an unreserved identifier R, and let:

$$\mathrm{R}(x, y, z) \Leftrightarrow (\exists w)\mathrm{PROD}(3, w, x) \vee [\mathrm{LESS}(y, x) \wedge \mathrm{LESS}(x, z)].$$

It follows that the desired relation is a generated one. What is more, the techniques outlined above allow us to write out an EFS procedure if we wish. Just this once, we go through the steps. It is simplest to break the problem down into elementary steps as follows. Let:

$$\mathrm{A}(w, x) \Leftrightarrow \mathrm{PROD}(3, w, x)$$
$$\mathrm{B}(x) \Leftrightarrow (\exists w)\mathrm{A}(w, x)$$
$$\mathrm{C}(x, y, z) \Leftrightarrow \mathrm{LESS}(y, x) \wedge \mathrm{LESS}(x, z)$$
$$\mathrm{D}(x, y, z) \Leftrightarrow \mathrm{B}(x) \vee \mathrm{C}(x, y, z).$$

It is easy to see that D and R are equivalent; the sequence above just decomposes the formula for R.

Now we can write the following sequence of procedures.

A(2):

PROD(3, w, x)→A(w, x)

B(1):

A(w, x)→B(x).

C(3):

LESS(y, x)→LESS(x, z)→C(x, y, z).

D(3):

B(x)→D(x, y, z);
C(x, y, z)→D(x, y, z).

The output of the procedure D is the desired relation.

Exercise 2.6.1. Assume that A, B, and C are reserved identifiers, and that D is unreserved. Let D(y, z)⇔(∃x)[A(x, y) ∨ B(x, z)] ∧ C(y, z). Write a procedure whose output is the relation defined by this formula.

Exercise 2.6.2. (trivial) Show that, for every EFS language, the equality relation is always generated.

Exercise 2.6.3. Work with EFS(**integer**) as given in §2.2. Use the methods of this section and show that the following is generated: the set of numbers y for which the equation $x^2 + 2xy = y^3$ has a solution in x. (For example, 3 is in the set because, when y is 3 the equation becomes $x^2 + 6x = 27$, which has a solution, namely 3.)

Exercise 2.6.4. Redo Exercise 2.5.1 using the methods of this section.

2.7 BOUNDED QUANTIFICATION

In the last section we introduced the *existential* quantifier, ∃, but not the *universal* quantifier, ∀, that usually goes with it. In logic, one informally reads (∀x)P(x) as: Every x meets the condition P. The difference in behavior between the two quantifiers, and the reason why we did not discuss ∀ in §2.6, needs some comment.

Say we are working in EFS(**integer**) and we have written a procedure R for a 2-place relation. Now consider the relation $S(x)$⇔(∃y)$R(x, y)$. How would we informally determine if, say, 3 is in S, that is, if (∃y)R(3, y)? We could do this by systematically doing computation after computation using R and looking through the outputs. If we ever find something of the form R(3, y) we know 3 is in S. If, after much searching, we have not found such an output, what does that tell us? Generally, nothing. We may not have found anything of the form R(3, y) because (1) nothing of this form can ever be an output, or (2) something of this form will turn up, but we haven't looked long enough yet. We can describe things using some terminology of more conventional programming languages: While $R(3, \text{something})$ not found, keep looking. This is a while loop with infinite looping possible if, in fact, nothing meets the condition $(\exists y)R(3, y)$. But if the condition is met, the loop must terminate.

The universal quantifier is quite a different creature. Suppose we set $T(x) \Leftrightarrow (\forall y)R(x, y)$. How would we check if 3 is in T? It would require us to verify that $R(3, 0)$ and $R(3, 1)$, and $R(3, 2)$, etc. There are infinitely many items to verify, and that is the sort of thing we should not expect machines to be generally capable of. In fact, in Chapter 6 we will *prove* that we do not always have closure of the generated relations under $\forall$. It does not happen in EFS(**integer**) for instance.

If we want to check whether $R(3, y)$ holds for all y in some *finite* set things are different. This is the sort of thing we expect machines to be able to do. In **integer** we can restrict a search to a finite set by saying we will only consider numbers within a certain range. In other structures we would say it differently. We now introduce a formal mechanism for this purpose, but it is structure dependent, so we work on a case-by-case basis. A certain amount of uniformity could have been introduced here if we had developed the machinery for recursive data structures, instead of just algebraic ones.

I. Integer

We introduce two bounded quantifiers, $(\forall y \leq z)$ and $(\exists y \leq z)$ as follows.

$(\exists y \leq z)R(x_1, \ldots, x_n, y)$ holds if $R(x_1, \ldots, x_n, y)$ holds for some value of y that is $\leq z$.

$(\forall y \leq z)R(x_1, \ldots, x_n, y)$ holds if $R(x_1, \ldots, x_n, y)$ holds for every value of y that is $\leq z$.

Proposition 2.7.1. *Suppose that R represents a generated $n+1$ place relation in EFS(**integer**), and we define S and T by*

$$S(x_1, \ldots, x_n, z) \Leftrightarrow (\exists y \leq z)R(x_1, \ldots, x_n, y)$$

$$T(x_1, \ldots, x_n, z) \Leftrightarrow (\forall y \leq z)R(x_1, \ldots, x_n, y).$$

Then both S and T are also generated.

Proof. S is easy because $(\exists y \leq z)\mathrm{R}(x_1, \ldots, x_n, y)$ is equivalent to $(\exists y)[\mathrm{LESS\_OR\_EQUAL}(y, z) \wedge \mathrm{R}(x_1, \ldots, x_n, y)]$. Then we apply the results of §2.6.

For T we write the following procedure (make sure you understand what it is doing).

T $(n+1)$:

$\mathrm{R}(x_1, \ldots, x_n, 0) \rightarrow \mathrm{T}(x_1, \ldots, x_n, 0);$

$\mathrm{T}(x_1, \ldots, x_n, w) \rightarrow \mathrm{SUC}(w, z) \rightarrow \mathrm{R}(x_1, \ldots, x_n, z) \rightarrow \mathrm{T}(x_1, \ldots, x_n, z).$

Closure of generated relations in EFS(**integer**) under bounded quantification makes it possible to establish easily that many nontrivial relations are generated. For example, an exercise in §2.2 showed that the set of nonprimes is generated; now we have a simple way of showing that the set of *primes* is also generated. Using relations from §2.2, let

PRIME(x) ⇔ NOT_EQUAL(x, 0) ∧ NOT_EQUAL(x, 1) ∧
($\forall y \leq x$)[EQUAL(y, 0) ∨ EQUAL(y, 1) ∨ EQUAL(y, x) ∨
NOT_DIVIDES(y, x)].

Exercise 2.7.1. Show the following is a generated relation of EFS(**integer**): y is the smallest prime bigger than x.

Exercise 2.7.2. Show using logical notation and bounded quantifiers that the following are generated relations of EFS(**integer**):

(a) z is the greatest common divisor of x and y.
(b) z is not the greatest common divisor of x and y.

Exercise 2.7.3. Suppose P represents a generated relation of EFS(**integer**), and NOT_P represents its complement, which is also generated. Let R(x, y) ⇔ x is the smallest integer such that P(x, y). Show that R is generated.

Exercise 2.7.4. Show the generated relations of EFS(**integer**) are closed under ($\forall x < y$) and ($\exists x < y$), where these bounded quantifiers have the obvious meanings.

II. str(L)

Here there are several natural notions of bounded quantifier to be considered. Recall, in Chapter 1 §1.5, the relation SUBWORD was determined to be generated (Exercise 1.5.3, part h). SUBWORD is transitive and reflexive on words, just as ≤ is on numbers, though it is not linear. Now we introduce the following.

($\exists y$ SUBWORD z)$R(x_1, \ldots, x_n, y)$ holds if $R(x_1, \ldots, x_n, y)$ holds for some value of y that is a subword of z.
($\forall y$ SUBWORD z)$R(x_1, \ldots, x_n, y)$ holds if $R(x_1, \ldots, x_n, y)$ holds for every value of y that is a subword of z.

Exercise 2.7.5. Show the generated relations of EFS(**str**(L)) are closed under ($\exists y$ SUBWORD z).

Proposition 2.7.2. *The generated relations of EFS*(**str**(L)) *are closed under* (∀y *SUBWORD* z).

Proof. The idea is this. Suppose we have a word with $k+1$ letters, say $a_1a_2\ldots a_ka_{k+1}$. If we have a subword that is not the whole word, it must omit the first or last letter (or both). Then it must be a subword of $a_2\ldots a_ka_{k+1}$ or of $a_1a_2\ldots a_k$, in either case, simpler (shorter) words. So, if we know every subword of $a_1a_2\ldots a_k$ has a certain property, and if we know every subword of $a_2\ldots a_ka_{k+1}$ has that property, and finally, if we know the word $a_1a_2\ldots a_ka_{k+1}$ itself has the property, it follows that every subword of $a_1a_2\ldots a_ka_{k+1}$ has the property.

Now, suppose that R represents a generated $n+1$ place relation of EFS(**str**(L)) and we define S by

$$\mathrm{S}(x_1, \ldots, x_n, z) \Leftrightarrow (\forall y \text{ SUBWORD } z)\mathrm{R}(x_1, \ldots, x_n, y).$$

The following procedure has the relation S as output. In it we use LETTER, given as Exercise 1.5.3(a). The first two statements take care of words with 0 and 1 letters. The third statement is for words of 2 or more letters. This procedure statement may be easier to understand if the reader keeps in mind the following scheme: z is the word awb, where a and b are variables intended to represent single letters; s is the "start" of z, aw; e is the "end" of z, wb.

S $(n+1)$:
 $\mathrm{R}(x_1, \ldots, x_n, \Delta) \rightarrow \mathrm{S}(x_1, \ldots, x_n, \Delta)$;
 $\mathrm{LETTER}(a) \rightarrow \mathrm{R}(x_1, \ldots, x_n, \Delta) \rightarrow \mathrm{R}(x_1, \ldots, x_n, a) \rightarrow \mathrm{S}(x_1, \ldots, x_n, a)$;
 $\mathrm{LETTER}(a) \rightarrow \mathrm{LETTER}(b) \rightarrow \mathrm{CON}(a, w, s) \rightarrow \mathrm{CON}(w, b, e) \rightarrow$
  $\mathrm{CON}(s, b, z) \rightarrow R(x_1, \ldots, x_n, z) \rightarrow \mathrm{S}(x_1, \ldots, x_n, s) \rightarrow$
  $\mathrm{S}(x_1, \ldots, x_n, e) \rightarrow \mathrm{S}(x_1, \ldots, x_n, z)$.

Exercise 2.7.6. Show the relation NOT_SUBWORD (with the obvious meaning) is generated.

Exercise 2.7.7. Show, by writing procedures, that the following are generated relations.

(a) BEGINS, word x is an initial subword of word y,
(b) ENDS, word x is a terminal subword of word y.

Exercise 2.7.8. Show the generated relations of EFS(**str**(L)) are closed under ($\forall x$ BEGINS y), ($\exists x$ BEGINS y), ($\forall x$ ENDS y), and ($\exists x$ ENDS y), where these bounded quantifiers have the obvious meanings.

Exercise 2.7.9. Show the generated relations of EFS(**str**(L)) are closed under ($\forall x$ SHORTER y) and ($\exists x$ SHORTER y), again with the obvious meanings.

III. tree(A), tree(A, $\neq$)

Exercise 2.7.10. Show the generated relations of EFS(**tree(A)**) and EFS(**tree(A,** $\neq$**)**) are closed under ($\forall x$ SUBTREE y) and ($\exists x$ SUBTREE y).

Exercise 2.7.11. Show the relation NOT_SUBTREE is a generated relation of EFS(**tree**(**A**, $\neq$)).

IV. set(A), set(A, $\neq$)

Exercise 2.7.12. Show the generated relations of EFS(**set**(**A**)) and EFS(**set**(**A**, $\neq$)) are closed under ($\forall x$ MEMBER_OF y) and ($\exists x$ MEMBER_OF y), with the obvious meanings.

Exercise 2.7.13. Show the generated relations of EFS(**set**(**A**)) and EFS(**set**(**A**, $\neq$)) are closed under ($\forall x$ SUBSET y) and ($\exists x$ SUBSET y). *Hint*: See Exercise 2.4.6.

2.8 MODELS AND PROGRAM CORRECTNESS

In Chapter 1, §1.6, the notions of *interpretation* and *model* were defined for the string manipulation languages considered there. Those definitions carry over essentially unchanged to the whole family of EFS languages. Remember, though, that an interpretation in a work space **W** assigns to identifiers sets and relations on the domain of **W**, which need not be words now. Still, Proposition 1.6.1 continues to hold, with no change in its proof.

Likewise, in Chapter 1, §1.7 the notion of *minimal* model was defined. That definition, too, carries over unchanged to all EFS languages. And Proposition 1.7.1, which was used to show the existence of minimal models, and Proposition 1.7.2, which said why we were interested in them, both carry over to all EFS languages with no changes in their proofs.

As an example, we sketch a proof of the correctness of the procedure NOT_MEMBER_OF from §2.4. We have a set **A** of atoms and we are working in an extension **W** of the basic work space of EFS(**set**(**A**, $\neq$)). The domain consists of all sets built up from members of **A**. The reserved identifiers are ADDMEM, representing the relation z is x with y added as member; ATOM, representing the set **A** of atoms; NOTEQ, representing the inequality relation on atoms; and MEMBER_OF, representing the $\in$ relation.

Define an interpretation **I** in the domain of work space **W** as follows. **I** assigns to ADDMEM, ATOM, NOTEQ, and MEMBER_OF the relations **W** specifies they represent. This ensures that **I** is an interpretation in **W**. Next, let **I** assign to DIFFERENT the relation $\neq$ on the domain of **W**, and to NOT_MEMBER_OF the relation $\notin$ on the domain of **W**. On other identifiers (which do not appear in the procedure NOT_MEMBER_OF) **I** is arbitrary. The claim is that **I** is a model for the procedure NOT_MEMBER_OF in **W**. We check one of the procedure statements:

NOT_MEMBER_OF(x, y)→DIFFERENT(x, w)
→ADDMEM(y, w, z)→NOT_MEMBER_OF(x, z).

Suppose we consider a substitution instance of this, replacing the variables x, y, z, and w by s_1, s_2, s_3, and s_4, which are members of the domain of **W**. We get

$$\text{NOT\_MEMBER\_OF}(s_1, s_2) \rightarrow \text{DIFFERENT}(s_1, s_4) \\ \rightarrow \text{ADDMEM}(s_2, s_4, s_3) \rightarrow \text{NOT\_MEMBER\_OF}(s_1, s_3).$$

We show this is true under interpretation **I**. If any of the conditions are false, we are done. Now suppose the conditions are all true; we must show the statement in the assignment position is also true. Since NOT_MEMBER_OF(s_1, s_2) is true under **I**, $\langle s_1, s_2 \rangle$ is in the relation $\text{NOT\_MEMBER\_OF}^{\mathbf{I}}$, which means that $s_1 \notin s_2$. Similarly, since DIFFERENT(s_1, s_4) is true under **I**, then $s_1 \neq s_4$. Finally, since ADDMEM(s_2, s_4, s_3) is true under **I**, then s_3 is s_2 with s_4 added. We have that s_1 is not in s_2, and it is not s_4, so s_1 cannot be in s_3; $s_1 \notin s_3$. Then, NOT_MEMBER_OF(s_1, s_3) is true under **I**.

We have shown that all subsitution instances of one of the procedure statements are true under **I**. The other procedure statements are treated similarly. It follows that **I** is a model for the procedure NOT_MEMBER_OF in **W**. Then by Proposition 1.6.1, the output of NOT_MEMBER_OF must be in the relation $\text{NOT\_MEMBER\_OF}^{\mathbf{I}}$. That is, if $\langle s_1, s_2 \rangle$ is in the output, then, in fact, $s_1 \notin s_2$.

Exercise 2.8.1. In §2.7, by writing a logical expression involving bounded quantifiers, the set of primes was shown to be generated in EFS(**integer**). Translate that into a procedure, and prove the procedure is correct, in the sense that its output does consist of only prime numbers.

2.9 PROCEDURE COMPLETENESS

To establish the completeness of a procedure we show the output includes everything we meant it to include. Completeness proofs are much less uniform than correctness proofs. They generally involve an induction on the complexity of the objects being manipulated, but that complexity is measured in different ways for different kinds of objects. We present two examples of completeness proofs, one simple, one more complicated.

Example 1. Consider the language EFS(**tree**(**A**)), and the procedure ATOM_OF, from §2.3. We show that if a is an atom occurring in tree t, then $\langle a, t \rangle$ is in the output of ATOM_OF.

Temporary Definition. By the *complexity* of a tree we mean the number of its branch points.

If t_1 and t_2 are left and right immediate subtrees of t, then the complexity of t is the complexity of t_1 + the complexity of $t_2 + 1$. Then, the complexity of any immediate subtree of t must be strictly less than the complexity of t.

We present an induction argument. The *induction hypothesis* is: For all trees t whose complexity is $<n$, if a is an atom of t, then $\langle a, t\rangle$ is an output of ATOM_OF. The *conclusion* we wish to establish is: If t is a tree whose complexity is exactly n, then $\langle a, t\rangle$ is an output of ATOM_OF whenever a is an atom of t. Now suppose t is a given tree, whose complexity is n.

Case 1: $n = 0$. Then t must be a trivial tree, consisting of just an atom, say a. Then $t = a$, and a is the only atom of t. But ATOM(a)$\rightarrow$ATOM_OF(a, a) is an instance of a procedure statement of ATOM_OF, and ATOM(a) is given in the work space. From these by the Assignment Rule we get ATOM_OF(a, a) or, since $t = a$, ATOM_OF(a, t). Thus, $\langle a, t\rangle$ is in the output, as desired.

Case 2: $n > 0$. Suppose that a is an atom of t. Since t is not trivial in this case, t has left and right immediate subtrees, say t_1 and t_2. If a is an atom of t, it must occur in t_1 or t_2, say in t_1. Now ATOM_OF(a, t_1)$\rightarrow$IMSUB(t, t_1, t_2)$\rightarrow$ATOM_OF(a, t) is an instance of a procedure statement of ATOM_OF. Also, t_1 is of lower complexity than t so by the induction hypothesis there is some trace T ending in ATOM_OF(a, t_1). Then the following is also a trace.

{trace T}
ATOM_OF(a, t_1)
IMSUB(t, t_1, t_2)
ATOM_OF(a, t_1)$\rightarrow$IMSUB(t, t_1, t_2)$\rightarrow$ATOM_OF(a, t)
ATOM_OF(a, t)

Hence $\langle a, t\rangle$ is a procedure output.

We have established the desired conclusion. Now, by induction, no matter what the complexity of t, if a is an atom of t, $\langle a, t\rangle$ is in the output of ATOM_OF.

Example 2. This time we use the language EFS(**set**(**A**, $\neq$)) and the procedure NOT_MEMBER_OF from §2.4. This is a more complicated example because we use two measures of complexity in the same argument: *depth* as defined in §2.4, and *size*, meaning the number of members. Two are needed because of the nature of the relation that ADDMEM represents: if $\langle s_1, s_2, s_3\rangle$ is in this relation, that is, if $s_3 = s_1 \cup \{s_2\}$, then the depth of s_2 is $<$ the depth of s_3, while the size of s_1 is $\leq$ the size of s_3.

Induction hypothesis: For all sets s whose depth is $<n$, if $t \notin s$, then $\langle t, s\rangle$ is in the output of NOT_MEMBER_OF.

To be shown: If s is a set of depth exactly n, then $\langle t, s\rangle$ is in the output of NOT_MEMBER_OF whenever $t \notin s$.

Suppose s is a set of depth n.

Case 1: $n = 0$. Then s is either an atom, or is the empty set.

Exercise 2.9.1. Give the argument necessary for case 1.

Case 2: $n > 0$. Suppose $t \notin s$; we show $\langle t, s \rangle$ is in the output of NOT_MEMBER_OF. We do this by an induction on the size of s, which we call k.

Subinduction hypothesis: For all sets s_0 of depth n and size $< k$, if $t \notin s_0$, then $\langle t, s_0 \rangle$ is in the output of NOT_MEMBER_OF.

To be shown: If s_1 is of depth n and size k, then $\langle t, s_1 \rangle$ is in the output of NOT_MEMBER_OF whenever $t \notin s_1$.

Suppose s_1 is a set of depth n and size k.

Subcase 2a: $k = 0$. Then s_1 must be an atom or the empty set, and the argument is as it was in case 1.

Subcase 2b: $k > 0$. Say $t \notin s_1$. We must show that $\langle t, s_1 \rangle$ is in the output of NOT_MEMBER_OF.

Since the size of s_1 is not 0, s_1 has members. Choose one, say s_2, and let s_0 be s_1 with s_2 removed. Then $s_1 = s_0 \cup \{s_2\}$, the depth of s_2 is smaller than that of s_1, which is n, and the size of s_0 is smaller than that of s_1, which is k. Also, since $t \notin s_1$, then $t \neq s_2$, and $t \notin s_0$.

Now, s_1 is s_0 with s_2 added, so ADDMEM(s_0, s_2, s_1) is given in the present work space. Also, s_0 and s_1 cannot be atoms, so SET(s_0) and SET(s_1) are given.

$t \notin s_0$, and the depth of s_0 must be $\leq n$. If the depth of s_0 is $< n$, by the *induction* hypothesis, $\langle t, s_0 \rangle$ is in the output of NOT_MEMBER_OF. If the depth $= n$, still the size of s_0 is $< k$, so by the *subinduction* hypothesis, $\langle t, s_0 \rangle$ is in the output of NOT_MEMBER_OF. Thus, no matter what, there is a trace ending in NOT_MEMBER_OF(t, s_0).

Also, $t \neq s_2$. This means one is an atom while the other isn't, or both are sets but one has a member which the other lacks. Say both are sets, and $s_3 \in t$ but $s_3 \notin s_2$; the other possibilities are handled similarly. Then SET(t) and SET(s_2) are given to us, as is MEMBER_OF(s_3, t). Further, since $s_3 \notin s_2$ and s_2 is of depth $< n$, by the *induction* hypothesis, NOT_MEMBER_OF(s_3, s_2) is the last line of a trace. Now,

$$\text{SET}(t) \rightarrow \text{SET}(s_2) \rightarrow \text{MEMBER\_OF}(s_3, t) \\ \rightarrow \text{NOT\_MEMBER\_OF}(s_3, s_2) \rightarrow \text{DIFFERENT}(t, s_2)$$

is an instance of a procedure statement of NOT_MEMBER_OF (the fourth). From all this it follows that there is a trace ending in DIFFERENT(t, s_2).

Finally,

$$\text{NOT\_MEMBER\_OF}(t, s_0) \rightarrow \text{DIFFERENT}(t, s_2) \rightarrow \text{ADDMEM}(s_0, s_2, s_1) \rightarrow \text{NOT\_MEMBER\_OF}(t, s_1)$$

is an instance of the last procedure statement. Since we have already verified all the conditions of this, the Assignment Rule gives us NOT_MEMBER_OF(t, s_1), which completes subcase 2b.

This finishes the "inner" induction, the subinduction. We have shown that for sets s of depth n, and any size, if $t \notin s$, then $\langle t, s \rangle$ is the output of NOT_MEMBER_OF. But this is exactly what was needed to complete the "outer" induction. The argument is finished.

Exercise 2.9.2. Work with EFS(**integer**) as defined in §2.2. Show the completeness of the procedures:

(a) SUM
(b) PROD
(c) LESS
(d) NOT_EQUAL

2.10 BACKGROUND

The family of EFS languages for arbitrary data structures has its origins in the brief notes [Smullyan 1956] and [Smullyan 1956A]. The ideas were further developed in [Fitting 1981], as part of a broader investigation that included notions not relevant to computability theory as such.

Like the EFS family, any style of programming language can be adapted to an arbitrary choice of data structure, and some recent programming languages actually allow the user considerable discretion in the creation of new data types.

The data structure used in logic programming is not one of those investigated in this chapter, but its description is simple. Suppose we have a finite collection **S** of *constant* and *function symbols,* with each function symbol having a specified *arity*. The set of *terms over* ***S***, **TM(S)**, is the smallest set containing the constant symbols of **S**, and containing $f(t_1, \ldots, t_n)$ whenever f is an n-place function symbol of **S** and $t_1, \ldots, t_n$ are terms over **S**. For each n-place function symbol f of **S** there is an $n + 1$ place *associated relation,* say **F**, where $\mathbf{F}(t_1, \ldots, t_n, t_{n+1})$ holds if t_{n+1} is the term $f(t_1, \ldots, t_n)$. The data structure **log(S)** is the one with domain **TM(S)** and with given relations those associated with the function symbols of **S**. Essentially, pure logic programming amounts to programming in EFS(**log(S)**) for various choices of **S**.

The language LISP essentially has **tree(A**, $\neq$) of §2.3 as its data structure. But [McCarthy 1967] is a quite readable discussion of the (unimplemented) family of languages having a LISP control structure, but allowing an arbitrary choice of data structure.

The use of **set**(**A**, $\neq$) as a data structure is less common in real-world computer usage than it might be. Pascal allows a very primitive application of a handful of ideas involving sets, but few other commonly used languages go even that far. At the other extreme, the experimental language SETL [Schwartz 1970], [Schwartz 1973] goes all the way, and takes sets as its primary data structure.

John von Neumann is best known today among computer people for his work on early computer design, the so-called von Neumann architecture. But earlier he did fundamental work in axiomatic set theory (as well as in many other areas of mathematics). In particular, the definition of ordinal number found in virtually all books on axiomatic set theory today is due to him [von Neumann 1923]. The sets chosen as number representatives in §2.4 are simply the finite von Neumann ordinals.

The relational data base example of §2.5 plays a minor role in this book. But relational, as opposed to hierarchical, data bases are taking on an increasingly important role in real-world computing. That a relational representation of data might have practical application was proposed in [Codd 1970] and further elaborated in [Codd 1972]. It should come as no surprise that PROLOG functions very well as a data-base query language. Again, we recommend [Kowalski 1979] for a readable discussion.

The material in §2.8 and §2.9 constitutes a program semantics for the family of EFS languages. Such semantics, together with the ability to use it to prove things about program behavior, is taking on a more central role in computing work as time goes on. [McCarthy 1967] contains interesting examples of proving results about LISP and LISP-like programs. [de Bakker 1980] is a thorough discussion of program semantics for a more conventional style of programming language. The reference section in de Bakker, with its discussion of the history of the subject (starting on page 466), is recommended.

REFERENCES

[Codd 1970] A relational model of data for large shared data banks, E. Codd, *Communications of the Association for Computing Machinery,* Vol. 13, pp. 377–387, 1970.

[Codd 1972] Relational completeness of data base sublanguages, E. Codd, *Data Base Systems,* R. Rustin, editor, Prentice-Hall, Englewood Cliffs, N.J., pp. 65–98, 1972.

[de Bakker 1980] *Mathematical Theory of Program Correctness,* J. de Bakker, Prentice-Hall International, Englewood Cliffs, N.J., 1980.

[Fitting 1981] *Fundamentals of Generalized Recursion Theory,* M. Fitting, North-Holland Publishing Co., Amsterdam, 1981.

[Kowalski 1979] *Logic for Problem Solving,* R. Kowalski, Elsevier-North Holland, New York, 1979.

[McCarthy 1967] A basis for a mathematical theory of computation, J. McCarthy, in *Computer Programming and Formal Systems,* P. Braffort and D. Hirschberg, editors, North Holland Publishing Co., Amsterdam, pp. 32–70, 1967.

[Schwartz 1970] *Set Theory as a Language for Program Specification and Programming,* J. Schwartz, Computer Science Department, Courant Institute of Mathematical Science, New York Universtiy, 1970.

[Schwartz 1973] *On Programming, An Interim Report on the SETL Project, Part I, Generalities, Part II, the SETL Language and Examples Of Its Use,* J. Schwartz, Courant Institute of Mathematical Science, New York University, 1973.

[Smullyan 1956] Elementary formal systems (abstract), R. Smullyan, *Bulletin of the American Mathematical Society,* Vol. 62, p. 600, 1956.

[Smullyan 1956A] On definability by recursion (abstract), R. Smullyan, *Bulletin of the American Mathematical Society,* Vol. 62, p. 601, 1956.

[von Neumann 1923] Zur Einführung der transfiniten Zahlen, *Acta litterarum ad scientiarum Regiae Universitatis Hungaricae Francisco-Josephinae,* Sectio scientiarum mathematicarum 1, pp. 199–208, 1923; translated as, On the introduction of transfinite numbers, in *From Frege to Gödel,* J. van Heijenoort editor, Harvard University Press, Cambridge, Mass., pp. 346–354, 1967.

3
OPERATORS

3.1 INTRODUCTION

Up to now, procedures generated outputs but there was no mechanism to supply them with inputs. In this chapter we provide one. We use the term *operator* for a mapping from relations to relations. Then, in this chapter, we add machinery to write procedures for operators.

Operators for which procedures can be written have a special *monotonicity* feature: If more information about the input is provided, no information about the output is lost. Monotone operators have pleasant properties that are mathematically useful for many purposes, so we begin with a discussion of monotone operators, before we consider how to write procedures for them. One can think of the general theory of monotone operators as a streamlining of the theory of mathematical inductions. As we have seen, proofs of procedure correctness always involve induction. A consequence of the work of this chapter is a unification of such proofs. There are other more significant consequences as well, but it is best to wait until some terminology has been introduced, and some preliminary work done, before trying to describe them.

3.2 MONOTONE OPERATORS

We begin the discussion of monotone operators by considering them in the abstract, postponing a discussion of their role in EFS languages. The results of the next several sections are very general, and the examples we present first have no direct connection with procedures.

Definition. Let $\mathbf{D}$ be some nonempty set. By an *operator* on the domain $\mathbf{D}$ we mean a function Φ that maps subsets of $\mathbf{D}$ to subsets of $\mathbf{D}$. An operator Φ on $\mathbf{D}$ is *monotone* if $A \subseteq B$ implies $\Phi(A) \subseteq \Phi(B)$ for any $A,\ B \subseteq \mathbf{D}$.

When discussing an operator, Φ, we will often refer to A as *input* and $\Phi(A)$ as *output.* In these terms, monotonicity says that if we increase the input from A to B, we don't lose anything that was part of the output. If $\mathbf{D}$ is a set of n-tuples then subsets of $\mathbf{D}$ are what we have been calling *relations.* So we are covering sets and relations with the definition above.

Examples. Our first examples have as domain **Z**, the set of integers (including the negative ones). Here are three operators on **Z**. For any $S \subseteq \mathbf{Z}$,

1. $\Phi_1(S) = \{x + 2 \mid x \in S\}$
2. $\Phi_2(S) = \{0\} \cup \{x + 2 \mid x \in S\}$
3. $\Phi_3(S) = \{0\} \cup \{x + 2 \mid x \in S\} \cup \{x - 2 \mid x \in S\}$

It is easy to check that these are monotone. If $S = \{1, 3\}$,

$\Phi_1(S) = \{3, 5\}$
$\Phi_2(S) = \{0, 3, 5\}$
$\Phi_3(S) = \{0, 3, 5, -1, 1\}$.

The next example has as domain the set of words over the alphabet $\{a, b\}$. It is closely related to the example discussed in Chapter 1, §1.7. For any $S \subseteq \{a, b\}^*$,

4. $\Phi_4(S) = \{a\} \cup \{xb \mid x \in S\}$

(where xb means the concatenation of x and b). If $S = \{b, ba\}$ then $\Phi_4(S) = \{a, bb, bab\}$.

The final example requires a small amount of terminology of group theory. If you are not familiar with it, skip the example. Nothing essential depends on it. Let **G** be some group, and let **H** be some fixed subset of **G**. Consider the operator such that, for any $S \subseteq \mathbf{G}$,

5. $\Phi_5(S) = \mathbf{H} \cup \{x^{-1} \mid x \in S\} \cup \{x \cdot y \mid x, y \in S\}$

In this "·" denotes the group operation. This, too, is easily seen to be monotone.

Definition. Let Φ be an operator on **D**. **F** is a *fixed point* of Φ if $\Phi(\mathbf{F}) = \mathbf{F}$. **F** is a *least* or *smallest* fixed point if **F** is a fixed point, and $\mathbf{F} \subseteq \mathbf{F}'$ for any fixed point $\mathbf{F}'$ of Φ. (It follows immediately that there can be only one least fixed point, provided there are any at all.)

For example, Φ_1, Φ_2, and Φ_3 all have **Z** as a fixed point. They also have the set **E** of all even integers (negative as well as positive) as a fixed point. In addition, Φ_1 has $\varnothing$ as a fixed point, and Φ_2 has the set of nonnegative even integers as a fixed point. In fact, the set of nonnegative even integers is the *least* fixed point for Φ_2. We verify this now.

By an induction argument, we show that if **F** is *any* fixed point of Φ_2, then $2 \cdot n \in \mathbf{F}$ for every nonnegative integer n. Well, $2 \cdot 0 = 0 \in \Phi_2(S)$ for any S, hence $0 \in \Phi_2(\mathbf{F}) = \mathbf{F}$ (**F** is a fixed point). Now suppose that $2 \cdot n \in \mathbf{F}$. By definition of Φ_2, $2 \cdot n + 2 \in \Phi_2(\mathbf{F}) = \mathbf{F}$. That is, $2 \cdot (n + 1) \in \mathbf{F}$. Then, by induction, every nonnegative even integer is in **F**. Since **F** was an arbitrary fixed point, and the set of nonnegative even integers itself is a fixed point, it must be the *least* one.

Exercise 3.2.1.

(a) Prove that the set of all even integers, positive and negative, is the least fixed point of Φ_3. What is the least fixed point of Φ_1?
(b) Determine several fixed points of Φ_4. What is the least one?

For another fixed point example, consider the operator Φ_5. If **H** is not empty, any fixed point of Φ_5 is a *subgroup* of **G**, extending **H**. This may be shown as follows. Let **F** be a fixed point, so $\Phi_5(\mathbf{F}) = \mathbf{F}$. Suppose $x \in \mathbf{F}$; then by definition $x^{-1} \in \Phi_5(\mathbf{F})$, hence $x^{-1} \in \mathbf{F}$. Thus, **F** is closed under inverses. Similarly, one shows that **F** is closed under the group multiplication operation, and that it extends **H**. This is enough to establish what we claimed.

Exercise 3.2.2. Prove a converse: Every subgroup of **G** that extends **H** is a fixed point of Φ_5.

The smallest subgroup of group **G** that extends **H** is the least fixed point of Φ_5 (its existence is a standard theorem of group theory, or follows by Theorem 3.3.1 in the next section). It is called the subgroup *generated by* **H**.

Not every operator has a fixed point. Consider, for example, the operator on **Z** given by

6. $\Phi_6(S) = \{x \in \mathbf{Z} \mid x \notin S\}$

Further, even if an operator has fixed points, it need not have a least one. Consider the operator on **Z** given by

7. $\Phi_7(S) = \begin{cases} S & \text{if } S \neq \varnothing \\ \mathbf{Z} & \text{if } S = \varnothing \end{cases}$

This has lots of fixed points, but none of them is a least one.

Exercise 3.2.3. Give an example of a monotone operator with exactly one fixed point; with exactly two fixed points; with exactly three fixed points.

Exercise 3.2.4. Suppose that Φ_a and Φ_b are two monotone operators on **D**. Union and intersection of operators can be defined as follows. For any $S \subseteq \mathbf{D}$, let

$$[\Phi_a \cup \Phi_b](S) = \Phi_a(S) \cup \Phi_b(S)$$
$$[\Phi_a \cap \Phi_b](S) = \Phi_a(S) \cap \Phi_b(S).$$

Prove that both $\Phi_a \cup \Phi_b$ and $\Phi_a \cap \Phi_b$ are monotone.

3.3 THE LEAST FIXED POINT THEOREM

The last section ended with examples of operators that did not have least fixed points. Those operators were not monotone. The main result of this section says why they could not have been.

Φ is monotone, $\Phi(\mathbf{M}_n) \subseteq \Phi(S)$. But $\Phi(\mathbf{M}_n) = \mathbf{M}_{n+1}$, and S is a semifixed point so $\Phi(S) \subseteq S$; hence $\mathbf{M}_{n+1} \subseteq S$. Then, by induction, $\mathbf{M}_n \subseteq S$ for every number n, and it follows that $\mathbf{M}_\omega \subseteq S$.

Exercise 3.4.1. Give a proof of Proposition 3.4.3 that does not use Proposition 3.4.2.

Exercise 3.4.2. Consider once again the operator Φ_2, given in §3.2. We have shown the least fixed point of it is the set of nonnegative even integers. This can also be verified using the machinery of this section.

(a) Show Φ_2 is compact.
(b) Let $\mathbf{M}_0, \mathbf{M}_1, \mathbf{M}_2, \ldots$ be the approximation sequence associated with Φ_2. Show by ordinary induction that $\mathbf{M}_{n+1} = \{0, 2, 4, 6, \ldots, 2n\}$. It follows that $\mathbf{M}_\omega$ is exactly the set of nonnegative even integers.

Exercise 3.4.3. Show that operator Φ_3 of §3.2 is compact, and say what its approximation sequence looks like.

Exercise 3.4.4. Give an example of an operator that is compact but not monotone.

Exercise 3.4.5. In Exercise 3.2.4 the union and the intersection of operators was defined and you showed the union and the intersection of monotone operators were again monotone. (See also Exercise 3.3.1.) Now show the union and the intersection of monotone, *compact* operators are compact.

Exercise 3.4.6. We will not be interested in operators that are compact but not monotone. Suppose the definition of compact had been replaced by: Φ is *strongly* compact provided $x \in \Phi(S)$ if and only if $x \in \Phi(S_0)$ for some finite $S_0 \subseteq S$.

(a) Show that if an operator is strongly compact, it must be monotone.
(b) Show that if an operator is compact and monotone, it must be strongly compact.

Exercise 3.4.7. Let Φ be a monotone, compact operator on $\mathbf{D}$. Suppose we have an increasing sequence, $S_0 \subseteq S_1 \subseteq S_2 \subseteq \ldots \subseteq \mathbf{D}$. Show $\Phi(\bigcup S_n) = \bigcup \Phi(S_n)$.

Exercise 3.4.8. Let Φ be an operator on $\mathbf{D}$, and suppose that for every increasing sequence, $S_0 \subseteq S_1 \subseteq S_2 \subseteq \ldots \subseteq \mathbf{D}$ we have $\Phi(\bigcup S_n) = \bigcup \Phi(S_n)$. Show Φ must be monotone, and if $\mathbf{D}$ is *countable* Φ must be compact.

3.5 MORE ON MONOTONE OPERATORS

In characterizing formal languages, such as programming or logic languages, two kinds of definitions are frequently encountered. One kind characterizes a class by saying how to construct its members. The other kind characterizes the class by saying it is the smallest set meeting certain closure conditions. It is generally possible to turn each kind of definition into the other. We illustrate this with a simple example, and present others as exercises. The methods of this section will be used in Chapter 5, §5.4.

The example we have chosen for illustration is the class of terms of a very simple formal language. We suppose a supply of *atomic* terms is available; for now we use $p, q, r, \ldots$ to stand for them. We have one *operation symbol,* $*$. Terms are intended to be things like $(p * q)$ and $((p * q) * (r * s))$ but not things like $(p(($ or $)(p * q$. We give two formal definitions of the class of terms, and we use a monotone operator to assist us in proving the two definitions are equivalent. These definitions will not be used after this section.

Term Definition 1. By a *formation sequence* we mean any finite sequence of words such that, for each w in the sequence either:

1. w is an atomic term, or
2. There are earlier words, u and v, in the sequence, and w is $(u * v)$.

Call a word a *type*-1-*term* if it is the last line of a formation sequence.

For example, assuming that p, q, r, and s are atomic terms, here is a formation sequence:

p
q
$(p * q)$
r
s
$(r * s)$
$((p * q) * (r * s))$

Thus $((p * q) * (r * s))$ is a type-1-term. In order to show something is a term according to this definition, we must say how to build it up, starting with atomic terms.

Exercise 3.5.1. Prove that $(p(($ is not a type-1-term.

If we put two formation sequences together into a longer sequence by first using all the words of one, in order, followed by all the words of the other, in order, the result is a new formation sequence. Also, if we take a formation sequence and make it shorter by throwing away words from the end, the result is still a formation sequence. These observations play a critical role later on.

Term Definition 2. Call a set S of words *closed* if:

1. S contains all atomic terms, and
2. Whenever $u, v \in S$ then $(u * v) \in S$.

Call a word a *type-2-term* if it is in the smallest closed set (we will establish there is such a set).

For example, again assume that p, q, r, and s are atomic terms. If S is *any* closed set, p and q are members by (1), hence $(p*q)$ is a member by (2). Similarly, $(r*s)$ is a member. But then, $((p*q)*(r*s))$ is a member, by (2) again. Since this is the case for any closed set, $((p*q)*(r*s))$ is in the smallest closed set (assuming there is one), and thus is a type-2-term.

We want to show that the class of type-1-terms and the class of type-2-terms are the same. As a tool we introduce an operator.

Let Φ be defined by $\Phi(S) = \{\text{atomic terms}\} \cup \{(u*v) \mid u, v \in S\}$.

You should verify that Φ is monotone and that the semifixed points of Φ are exactly the sets called closed above. You might also verify that Φ is compact, though we will not need this here. Since monotone operators have least fixed points there is a smallest closed set, hence the definition of type-2-term is meaningful. Our job has become: show the members of the least fixed point of Φ are exactly the type-1-terms.

Exercise 3.5.2. Prove that $(p(($ is not a type-2-term.

Proof, part 1. We show every type-1-term is in *any* fixed point S of Φ, hence is in the least one, and so is a type-2-term. Our argument is by induction on the lengths of formation sequences. Let S be some fixed point of Φ.

Induction hypothesis: Whenever we have a formation sequence of length $< n$, its last line is in S.

To be shown: The last line of a formation sequence whose length is exactly n, is in S.

Suppose we have a formation sequence $w_1, w_2, \ldots, w_n$ of length n. We show its last word, w_n, is in S. By definition, there are two possibilities. Case 1: w_n is an atomic term. Then $w_n \in \Phi(S)$ by definition of Φ, hence $w_n \in S$ since S is a fixed point. Case 2: There are earlier words w_i and w_j in the sequence, and w_n is $(w_i * w_j)$. But then, $w_1, \ldots, w_i$ and $w_1, \ldots, w_j$, being beginning parts of a formation sequence, are themselves formation sequences, and of length $< n$. By the induction hypothesis, both w_i and w_j are in S. It follows that $(w_i * w_j)$, that is w_n, is in $\Phi(S)$ and hence in S. In either case, $w_n \in S$. Then by induction, every type-1-term is in S.

Proof, part 2. We show that every member of the least fixed point of Φ, that is, every type-2-term, is a type-1-term. This time we use generalized induction, Theorem 3.3.3.

Let P be the set of type-1-terms; we show that $\Phi(P) \subseteq P$. Well, suppose $w \in \Phi(P) = \{\text{atomic terms}\} \cup \{(u*v) \mid u, v \in P\}$; we show that $w \in P$. There are two possibilities again. Case 1: $w \in \{\text{atomic terms}\}$. Then consider the one-line sequence whose only word is w. This is a formation sequence whose last line is w. Hence, w is a type-1-term; $w \in P$. Case 2: $w \in \{(u * v) \mid u, v \in P\}$. So there are u, $v \in P$ with $w = (u * v)$. Then there are two formation sequences, one ending with u and one ending with v: $s_1, \ldots, s_i, u$ and $t_1, \ldots, t_j, v$. As we observed earlier, if we put them together into a single sequence $s_1, \ldots, s_i, u, t_1, \ldots, t_j, v$ this is still a formation sequence. And, if we add $(u * v)$ as a new last item, $s_1, \ldots, s_i, u, t_1, \ldots, t_j, v, (u * v)$, we still have a formation sequence. Thus, $(u * v)$, or w, is again a type-1-term; $w \in P$. We have shown that $\Phi(P) \subseteq P$. It follows that the least fixed point of Φ is a subset of P.

Exercise 3.5.3. In Chapter 1, §1.3 we defined the class of statements of an EFS language. Give another definition analogous to the one above using formation sequences, assuming a definition of atomic statement has already been given. Also give a definition involving an analog of the closure conditions above. Prove the two definitions equivalent.

In Chapter 1, §1.4 we defined the notion of a restricted trace. There are obvious similarities between restricted traces and formation sequences, which suggest the following.

Definition. Let NAME be a procedure in work space **W**. By a *derivation set* for NAME in **W** we mean a set S of procedure statements such that:

1. S contains every instance of the given relations of **W**.
2. S contains every substitution instance of the procedure statements of NAME.
3. If X and $X \rightarrow F$ are in S, where X is atomic, then F is in S.

Proposition 3.5.1. *IDENTIFIER*$(\mathrm{t}_1, \ldots, \mathrm{t}_j)$ *is the last line of a restricted trace of NAME in work space* **W** *if and only if IDENTIFIER*$(\mathrm{t}_1, \ldots, \mathrm{t}_j)$ *is a member of the smallest derivation set for NAME in* **W**. *Hence* $\langle \mathrm{t}_1, \ldots, \mathrm{t}_n \rangle$ *is in the output of NAME exactly when NAME*$(\mathrm{t}_1, \ldots, \mathrm{t}_n)$ *is in the smallest derivation set for NAME in* **W**.

Exercise 3.5.4. Prove Proposition 3.5.1.

3.6 PROCEDURES THAT ACCEPT INPUT

It is time to return to EFS languages and bring operators into the picture. We begin with *syntax*. The header of a procedure has specified the output identifier. From now on, we also allow an input identifier to be specified.

Definition. Suppose IN_NAME and OUT_NAME are distinct identifiers, and n and k are positive integers. Then the following is also a procedure *header:*

OUT_NAME (n) INPUT IN_NAME (k):

If OUT_NAME (n) INPUT IN_NAME (k): is the header of a procedure, we use OUT_NAME as a name for the procedure, as well as being the identifier representing the output (which is an n-place relation). IN_NAME is intended to represent input to the procedure, a k-place relation.

As usual, procedures are *in* a work space. The general definitions and requirements are the same as in earlier chapters but with the following also.

Conditions. In addition to the conditions of earlier chapters. In the *body* of a procedure with header OUT_NAME (n) INPUT IN_NAME (k): The identifier IN_NAME may not occur in the assignment position of any procedure statement. For a procedure with header OUT_NAME (n) INPUT IN_NAME (k): to be *in* work space **W** the identifier IN_NAME must not be a reserved identifier of **W**.

The intention behind these extra requirements is simple. If the identifier representing input were reserved in the work space, it would represent a fixed, given relation. A procedure that accepts input is not very interesting if the input can't be changed. Also, if the identifier representing input occurred in the assignment position of some procedure statement, one could assign values to it in the course of a computation. This is contrary to the intuition of an input, which is something we are given "from the outside," not something we compute.

Now that we have said what procedures that accept input "look like," we say how they behave. We give their *operational semantics*. We begin with a background discussion, leading to the notion of an augmented trace.

Suppose **W** is a work space, and the following is a procedure in **W**.

OUT_NAME (n) INPUT IN_NAME (k):
  (procedure body of OUT_NAME).

Let **P** be some k-place relation on the domain of **W**. We say what the output of OUT_NAME is to be when **P** is used as input. Let **W**′ be a work space that is like **W** except: (1) IN_NAME is added to the list of reserved identifiers, and (2) IN_NAME is specified as representing **P**. Since IN_NAME was not a reserved identifier of **W**, it makes sense to talk about adding it to the list of reserved identifiers. Also, and this is most important, every procedure statement of OUT_NAME is still acceptable in **W**′, because we required that IN_NAME should not occur in assignment positions. So the following is a procedure in work space **W**′, of the sort we have been considering in earlier chapters.

OUT_NAME (n):
  (procedure body of OUT_NAME).

We take the output of OUT_NAME in work space **W**, with **P** as input, to be the output of the no-input procedure above, having the same procedure body, but used in work space **W**′. We denote this output by OUT_NAME(**P**). This means that to determine the members of OUT_NAME(**P**) in **W**, we must run traces of OUT_NAME in **W**′, but these are just traces as we have known them for the last two chapters. On the other hand, the use of a different work space for each different input relation is awkward.

Definition. Suppose we have a procedure in work space **W** with header OUT_NAME (n) INPUT IN_NAME (k):, and suppose **P** is a k-place relation on the domain of **W**. By an *augmented trace* of OUT_NAME(**P**) we mean a trace of OUT_NAME in **W** itself, allowing the extra rule: IN_NAME($t_1, \ldots, t_k$) can be used as a line whenever $\langle t_1, \ldots, t_k \rangle$ is in the relation **P**. $\langle u_1, \ldots, u_n \rangle$ is in OUT_NAME(**P**) if there is an augmented trace of OUT_NAME(**P**) whose last line is OUT_NAME($u_1, \ldots, u_n$).

An augmented trace of OUT_NAME(**P**) in **W** is really a trace of OUT_NAME in **W**′. From now on, augmented traces are all we display, and work spaces other than **W** itself will no longer be mentioned. For example, let L be the one-letter alphabet $\{1\}$ and consider the basic work space of EFS(**str**(L)). The following procedure is in this work space.

OUT (1) INPUT IN (1):
  OUT(Δ);
  IN(x)→CON(x, 11, y)→OUT(y).

Here is an augmented trace of OUT($\{1, 111\}$) in the basic work space of EFS(**str**(L)).

IN(1)
CON(1, 11, 111)
IN(1)→CON(1, 11, 111)→OUT(111)
OUT(111).

Thus, 111 is in the output of OUT, using $\{1, 111\}$ as input. That is, $111 \in$ OUT($\{1, 111\}$).

Exercise 3.6.1. What are

(a) OUT(∅)
(b) OUT(OUT($\{1, 111\}$))?

Exercise 3.6.2. Let $L = \{a, b, c\}$ and use the basic work space of EFS(**str**(L)). Write an input-accepting procedure NOC such that, if **P** is any set of words over L, then NOC(**P**) is the subset of **P** made up of words not containing c.

Exercise 3.6.3. Show that every generated relation is the output of a constant input-accepting procedure.

Exercise 3.6.4. Let **W** be a work space. Suppose f is a function on the domain of **W**, and suppose the graph of f is generated. Show there is an input-accepting procedure F in **W** such that, for each w, $F(\{w\}) = \{f(w)\}$.

Remark. For EFS(**str**(L)), EFS(**integer**), and a variety of other EFS languages, the converse of this exercise is also true: if there is an input-accepting procedure F such that $F(\{w\}) = \{f(w)\}$, then the graph of f is a generated relation. A proof of this would take us too far afield here however.

3.7 PROCEDURES AS OPERATORS

Suppose **W** is a work space, and we write an input-accepting procedure in **W**, say

OUT_NAME (n) INPUT IN_NAME (n):
(procedure body of OUT_NAME).

OUT_NAME, when supplied with an n-place relation as input, yields an n-place relation as output. Thus, we have a mapping from relations to relations, what we have been calling an *operator*. More precisely, it is an operator on $\mathbf{D}^n$, where **D** is the domain of work space **W**, and $\mathbf{D}^n$ is the set of all n-tuples of members of **D**. Now we apply the material of earlier sections to such operators.

We have restricted things to the case where both input and output are n-place relations. The definition of procedure allows input to be k-place and output n-place, where n and k are different. Much of what we develop in this section applies in the more general case as well, and for the same reasons. But we will have no need of it. Our principal interest is in fixed points of operators, and the whole issue makes no sense if outputs and inputs are different-sized relations.

Definition. Let the following be a procedure in work space **W**, with domain **D**.

OUT_NAME (n) INPUT IN_NAME (n):
(procedure body of OUT_NAME).

By the *operator defined by* this procedure we mean the operator Φ on $\mathbf{D}^n$ such that, for any $\mathbf{P} \subseteq \mathbf{D}^n$, $\Phi(\mathbf{P}) = \text{OUT\_NAME}(\mathbf{P})$. That is, $\Phi(\mathbf{P})$ is the output of OUT_NAME when **P** is used as input. We generally use OUT_NAME to denote this operator, as well as being the name for the procedure that defines it.

Proposition 3.7.1. *An operator defined by a procedure is monotone.*

Proof. Suppose

OUT_NAME (n) INPUT IN_NAME (n):
(procedure body of OUT_NAME).

is a procedure in **W**, the domain of **W** is **D**, **P** and **Q** are both subsets of $\mathbf{D}^n$ and $\mathbf{P} \subseteq \mathbf{Q}$. It is easy to see that any augmented trace of OUT_NAME(**P**) in **W** is also an augmented trace of OUT_NAME(**Q**) in **W**. It follows that if OUT_NAME($u_1, \ldots, u_n$) is the last line of an augmented trace of OUT_NAME(**P**), it is also the last line of an augmented trace of OUT_NAME(**Q**), so OUT_NAME(**P**) $\subseteq$ OUT_NAME(**Q**).

Then Theorems 3.3.2 and 3.3.3 apply: Operators defined by procedures have least fixed points, and we can prove things about these least fixed points using generalized induction.

Proposition 3.7.2. *An operator defined by a procedure is compact.*

Proof. Suppose

OUT_NAME (n) INPUT IN_NAME (n):
    (procedure body of OUT_NAME).

is a procedure in **W**, the domain of **W** is **D**, **P** is a subset of $\mathbf{D}^n$, and $\langle t_1, \ldots, t_n \rangle \in$ OUT_NAME(**P**). Then there is an augmented trace T of OUT_NAME(**P**) in **W** with OUT_NAME($t_1, \ldots, t_n$) as last line. T, of course, has only finitely many lines. Let $\mathbf{P}_0$ be the set of all $\langle u_1, \ldots, u_n \rangle$ such that IN_NAME($u_1, \ldots, u_n$) actually occurs as a line of T. Then $\mathbf{P}_0$ will be a finite set, and obviously T is also an augmented trace of OUT_NAME($\mathbf{P}_0$). But if IN_NAME($u_1, \ldots, u_n$) occurs as a line of T it must be that $\langle u_1, \ldots, u_n \rangle \in \mathbf{P}$ (recall IN_NAME could not be used in the assignment position of any procedure statement, nor was it reserved in **W**). It follows that $\mathbf{P}_0 \subseteq \mathbf{P}$. Thus, $\langle t_1, \ldots, t_n \rangle \in$ OUT_NAME($\mathbf{P}_0$) for a finite subset $\mathbf{P}_0$ of **P**.

This is a manifestation of the finitary aspect of computers. For each particular procedure output, only a finite portion of input can have been used in generating that output. Since operators defined by procedures are compact, Proposition 3.4.3 applies. Not only do least fixed points exist, but we have a "natural" mechanism for approximating to them.

Exercise 3.7.1. Consider the following procedure in the basic work space of EFS(**str**({1})), introduced in §3.6.

OUT (1) INPUT IN (1):
    OUT(Δ);
    IN(x)→CON(x, 11, y)→OUT(y).

Determine the approximation sequence associated with the operator this procedure defines. And, hence, determine the least fixed point of the operator.

Exercise 3.7.2. Consider the following procedure in the basic work space of EFS(**str**({a, b})).

O (2) INPUT I (2):
    O(Δ, Δ);
    I(u, v)→CON(x, y, u)→CON(x, a, t)→CON(t, y, w)→CON(v, a, z)→
        O(w, z);
    I(u, v)→CON(x, y, u)→CON(x, b, t)→CON(t, y, w)→CON(v, b, z)→
        O(w, z).

(a) Determine the first four terms of the approximation sequence associated with the operator O.

(b) Use the Generalized Induction Theorem 3.3.3 to show that if $\langle t_1, t_2 \rangle$ is in the least fixed point of the operator O, then t_1 and t_2 contain the same number of a's, and the same number of b's.

(c) Show that if t_1 and t_2 are words over $\{a, b\}$ that contain the same number of a's and the same number of b's, then $\langle t_1, t_2 \rangle$ is in the least fixed point of the operator O. *Hint*: Use induction on the length of t_2 (which must be the same as the length of t_1).

Exercise 3.7.3. In Exercise 3.2.4 the union and the intersection of operators were defined. (Note Exercises 3.3.1 and 3.4.5 also.) Suppose we have the following two procedures in work space **W**.

OUT_A (n) INPUT IN_A (n):
  (procedure body A).

and

OUT_B (n) INPUT IN_B (n):
  (procedure body B).

Both of these define operators, for which we can form the union and intersection.

(a) Write a procedure that defines the union of the OUT_A operator and the OUT_B operator.

(b) Write a procedure that defines the intersection of the OUT_A operator and the OUT_B operator.

3.8 DENOTATIONAL SEMANTICS

In Chapter 1, §1.6 and Chapter 2, §2.8, we defined *models* for procedures that don't accept input. Now we want to take inputs into account. In §3.6 we said an input-accepting procedure, when given **P** as input, behaved like a closely related procedure that did not accept input, in a work space expansion with **P** as an added given relation. Since this reduces things to the behavior of ordinary procedures, the semantics of earlier chapters applies. But this notion of model is somewhat awkward to work with. So we start over, define the models needed for input-accepting procedures in a different way and at the end show we wind up in the same place.

Definition. Suppose that $\mathbf{P} \subseteq \mathbf{D}^k$ and the following is a procedure in work space **W**, with domain **D**.

OUT_NAME (n) INPUT IN_NAME (k):
  (procedure body of OUT_NAME).

We call **M** a ***P** input model* for the procedure OUT_NAME in **W** if (1) **M** is a model of OUT_NAME in work space **W** as defined in earlier chapters, and (2) $\mathbf{P} \subseteq$ IN_NAME$^{\mathbf{M}}$.

For example, consider again the procedure

OUT (1) INPUT IN (1):
OUT(Δ);
IN(x)→CON(x, 11, y)→OUT(y).

And let $\mathbf{P} = \{1, 111\}$. If you did Exercise 3.6.1 you verified that OUT(**P**) = $\{\Delta, 111, 11111\}$. Now consider the following interpretations.

1. $\mathbf{I}_1$ assigns to IN the set $\{1, 111\}$
   to OUT the set $\{\Delta, 111, 11111\}$
   to CON the relation **CON**.
2. $\mathbf{I}_2$ assigns to IN the set $\{1, 11, 111\}$
   to OUT the set $\{\Delta, 111, 11111\}$
   to CON the relation **CON**.
3. $\mathbf{I}_3$ assigns to IN the set $\{1, 11, 111\}$
   to OUT the set $\{\Delta, 111, 1111, 11111\}$
   to CON the relation **CON**.

All three are interpretations in the basic work space of EFS(**str**({1})). And all three meet the condition $\mathbf{P} \subseteq \mathrm{IN}^{\mathbf{M}}$. Further, $\mathbf{I}_1$ and $\mathbf{I}_3$ are models for the procedure OUT, but $\mathbf{I}_2$ is not. So $\mathbf{I}_1$ and $\mathbf{I}_3$ are ***P** input models for OUT,* but $\mathbf{I}_2$ is not.

Exercise 3.8.1. Suppose that **M** is a **P** input model in **W** for a procedure OUT_NAME. Show that, if $\langle t_1, \ldots, t_n \rangle \in$ OUT_NAME(**P**) then $\langle t_1, \ldots, t_n \rangle$ is in the relation OUT_NAME$^{\mathbf{M}}$. *Hint:* Use induction on the length of an augmented trace.

In Chapter 1, §1.7 we defined the notion of intersection for a family of interpretations. It is easy to see that, if for every member **I** of the family $\{\mathbf{I}_1, \mathbf{I}_2, \mathbf{I}_3 \ldots\}$ we have $\mathbf{P} \subseteq$ IN_NAME$^{\mathbf{I}}$, and if $\mathbf{J} = \bigcap\{\mathbf{I}_1, \mathbf{I}_2, \mathbf{I}_3 \ldots\}$, then $\mathbf{P} \subseteq$ IN_NAME$^{\mathbf{J}}$. Then, applying Proposition 1.7.1, we have the following.

*The intersection of a family of **P** input models for OUT_NAME in **W** is also a **P** input model for OUT_NAME in **W**.*

Exercise 3.8.2. Show that there is always at least one **P** input model for OUT_NAME in **W**.

Definition. The *minimal* **P** input model for OUT_NAME in **W** is the intersection of all **P** input models for OUT_NAME in **W**. (Existence and uniqueness are straightforward, and by the remarks above, it must itself be a **P** input model for OUT_NAME.)

We have allowed interpretations of the input identifier in **P** input models to be bigger than **P**. By not insisting on having **P** exactly, models become easier to construct, since we do not have such rigid conditions to meet. The following says this extra freedom makes no difference when we come to the minimal model.

Proposition 3.8.1. *Suppose that* $\mathbf{P} \subseteq \mathbf{D}^k$ *and the following is a procedure in work space* **W**, *with domain* **D**.

OUT_NAME (n) *INPUT IN_NAME* (k):
(*procedure body of OUT_NAME*).

Let **M** *be the minimal* **P** *input model for OUT_NAME in* **W**. *Then* $IN\_NAME^{\mathbf{M}} = \mathbf{P}$.

Proof. Since **M** is a **P** input model for OUT_NAME, we have $\mathbf{P} \subseteq$ IN_NAME$^{\mathbf{M}}$. Suppose **P** were a *proper* subset of IN_NAME$^{\mathbf{M}}$. Say $\langle a_1, \ldots, a_k \rangle \in$ IN_NAME$^{\mathbf{M}}$ but $\langle a_1, \ldots, a_k \rangle \notin \mathbf{P}$. Define an interpretation **I** as follows. On all identifiers except IN_NAME, **I** and **M** are the same. And IN_NAME$^{\mathbf{I}}$ is to be IN_NAME$^{\mathbf{M}}$ with $\langle a_1, \ldots, a_k \rangle$ removed. That is, IN_NAME$^{\mathbf{I}}$ = IN_NAME$^{\mathbf{M}}$ − $\{\langle a_1, \ldots, a_k \rangle\}$.

M is an interpretation in **W**, but IN_NAME is not a reserved identifier of **W**, hence **I** and **M** agree on the reserved identifiers, and **I** is an interpretation in **W**. $\mathbf{P} \subseteq$ IN_NAME$^{\mathbf{M}}$, and $\langle a_1, \ldots, a_k \rangle \notin \mathbf{P}$, hence $\mathbf{P} \subseteq$ IN_NAME$^{\mathbf{M}}$ − $\{\langle a_1, \ldots, a_k \rangle\}$. That is, $\mathbf{P} \subseteq$ IN_NAME$^{\mathbf{I}}$; **I** is a **P**-input model. Suppose $X_1 \to X_2 \to \ldots \to X_p \to T$ is a substitution instance of a procedure statement of OUT_NAME; we show it is true under **I**. Well, suppose each of $X_1, X_2, \ldots, X_p$ is true under **I**; we show T is also true under **I**. Since each X_i is true under **I**, and every atomic statement that **I** calls true, **M** calls true, each X_i is true under **M**. Since **M** is a model for OUT_NAME, $X_1 \to X_2 \to \ldots \to X_p \to T$ is true under **M** and it follows that T is true under **M**. T is of the form IDENTIFIER($t_1, \ldots, t_m$) but, since IN_NAME was not allowed to occur in assignment positions, IDENTIFIER must be different than IN_NAME. Hence **I** and **M** agree on IDENTIFIER, so T must be true under **I** as well.

We have just verified that **I** itself is a **P** input model for OUT_NAME in **W**. But **M** was defined to be the *intersection* of all such models; $\langle a_1, \ldots, a_k \rangle \in$ IN_NAME$^{\mathbf{M}}$ but $\langle a_1, \ldots, a_k \rangle \notin$ IN_NAME$^{\mathbf{I}}$. This is impossible. Hence **P** could not have been a proper subset of IN_NAME$^{\mathbf{M}}$. This concludes the proof.

Now the result that says **P** input models and the models of §3.6 yield the same semantics.

Proposition 3.8.2. *Once again, suppose that* $\mathbf{P} \subseteq \mathbf{D}^k$ *and the following is a procedure in work space* **W**, *with domain* **D**.

OUT_NAME (n) *INPUT IN_NAME* (k):
(*procedure body of OUT_NAME*).

*Let **M** be the minimal **P** input model for OUT_NAME in **W**. Further, let **W**′ be the work space that results when IN_NAME is added to the list of reserved identifiers of **W**, and is specified as representing **P**. Let **M**′ be the minimal model (as defined in earlier chapters) for the following no-input procedure in **W**′.*

OUT_NAME (n):
(procedure body of OUT_NAME).

*Then **M** = **M**′; that is, **M** and **M**′ assign the same relations to each identifier.*

Proof. **M**′ meets the conditions for being a **P** input model for OUT_NAME in **W**. **M** is the minimal such model. Hence, for each identifier, we have $\text{IDENTIFIER}^{\mathbf{M}} \subseteq \text{IDENTIFIER}^{\mathbf{M}'}$. By Proposition 3.8.1, **M** is a model for the no-input procedure OUT_NAME in work space **W**′. **M**′ is the minimal such model. Hence $\text{IDENTIFIER}^{\mathbf{M}'} \subseteq \text{INDENTIFIER}^{\mathbf{M}}$ for every identifier. Thus, **M** = **M**′.

Corollary 3.8.3. *OUT_NAME(**P**) is the interpretation of OUT_NAME in the minimal **P** input model for OUT_NAME in **W**.*

Proof. By Proposition 3.8.2 and the characterization of OUT_NAME(**P**) given in §3.6.

We sum things up. For a *k*-place relation **P**, let **F**(**P**) be the family of all **P** input models for OUT_NAME in work space **W**. Then:

1. **F**(**P**) is not empty.
2. $\bigcap$**F**(**P**) is itself a **P** input model, the minimal one.
3. $\text{IN\_NAME}^{\cap \mathbf{F}(\mathbf{P})} = \mathbf{P}$.
4. $\text{OUT\_NAME}^{\cap \mathbf{F}(\mathbf{P})} = \text{OUT\_NAME}(\mathbf{P})$.

3.9 MORE ON MODELS

In §3.7, results were established that, though simple, are fundamental in what follows. Propositions 3.7.1 and 3.7.2 were proved using augmented traces as tools. It is also possible to prove them using models. We believe doing so provides additional insight into the results themselves and into the behavior of models. In this section, to keep language clutter down we assume we are dealing with a fixed EFS language and a single work space, **W**. All interpretations mentioned are of that language, in that work space.

Definition. Let **I** and **J** be two interpretations. We write $\mathbf{I} \leq \mathbf{J}$ to mean $\text{IDENTIFIER}^{\mathbf{I}} \subseteq \text{IDENTIFIER}^{\mathbf{J}}$ for every identifier of the language.

For example, in Chapter 1, §1.7 several interpretations were defined, and their ≤ ordering was presented in a diagram. Things were simple there because the ordering only depended on the interpretation of a single identifier. In §3.8 of this chapter, three interpretations were defined. You should check that, for these, $\mathbf{I}_1 \le \mathbf{I}_2 \le \mathbf{I}_3$.

Exercise 3.9.1. Show that $\le$ is transitive, reflexive, and antisymmetric. That is:

(a) $\mathbf{I} \le \mathbf{J}$ and $\mathbf{J} \le \mathbf{K}$ imply $\mathbf{I} \le \mathbf{K}$.
(b) $\mathbf{I} \le \mathbf{I}$.
(c) $\mathbf{I} \le \mathbf{J}$ and $\mathbf{J} \le \mathbf{I}$ imply $\mathbf{I} = \mathbf{J}$.

Exercise 3.9.2. Show that for an *atomic* procedure statement X, with or without variables, if $\mathbf{I} \le \mathbf{J}$ and X is true under $\mathbf{I}$, then X is true under $\mathbf{J}$. (This is not true in general for nonatomic procedure statements, as the example on unions later in this section shows.)

Exercise 3.9.3. Let $\mathbf{F}$ be a family of interpretations, and suppose $\mathbf{I} \in \mathbf{F}$. Show that $\bigcap \mathbf{F} \le \mathbf{I}$.

Exercise 3.9.4. Let OUT_NAME be an input-accepting procedure and let $\mathbf{M}$ be a $\mathbf{P}$ input model for OUT_NAME. Show $\mathbf{M}$ is the *minimal* $\mathbf{P}$ input model for OUT_NAME if and only if $\mathbf{M} \le \mathbf{I}$ for all $\mathbf{P}$ input models $\mathbf{I}$ for OUT_NAME.

Exercise 3.9.5. Let $\mathbf{F}$ and $\mathbf{G}$ be two nonempty families of interpretations, and suppose that $\mathbf{F} \subseteq \mathbf{G}$. Show that $\bigcap \mathbf{G} \le \bigcap \mathbf{F}$. Use this to redo Exercise 3.9.3.

Proposition 3.9.1. *An operator defined by a procedure is monotone. (This is Proposition 3.7.1 again, but now we give a semantic proof.)*

Proof. Consider the following procedure in work space $\mathbf{W}$.

OUT_NAME (n) INPUT IN_NAME (n):
  (procedure body of OUT_NAME).

Let $\mathbf{P}$ and $\mathbf{Q}$ be n-place relations on the domain of $\mathbf{W}$, and suppose $\mathbf{P} \subseteq \mathbf{Q}$.

Suppose $\mathbf{M}$ is a $\mathbf{Q}$ input model for OUT_NAME in $\mathbf{W}$. Then $\mathbf{Q} \subseteq \text{IN\_NAME}^{\mathbf{M}}$. Since $\mathbf{P} \subseteq \mathbf{Q}$, $\mathbf{P} \subseteq \text{IN\_NAME}^{\mathbf{M}}$, so $\mathbf{M}$ is also a $\mathbf{P}$ input model. Let $\mathbf{F}(\mathbf{P})$ be the family of all $\mathbf{P}$ input and $\mathbf{F}(\mathbf{Q})$ the family of all $\mathbf{Q}$ input models for OUT_NAME in $\mathbf{W}$. What we just showed is $\mathbf{F}(\mathbf{Q}) \subseteq \mathbf{F}(\mathbf{P})$. Then by Exercise 3.9.5, $\bigcap \mathbf{F}(\mathbf{P}) \le \bigcap \mathbf{F}(\mathbf{Q})$. And so for each identifier, $\text{IDENTIFIER}^{\cap \mathbf{F}(\mathbf{P})} \subseteq \text{IDENTIFIER}^{\cap \mathbf{F}(\mathbf{Q})}$.

By Corollary 3.8.3, OUT_NAME($\mathbf{P}$) is the interpretation of OUT_NAME in the minimal $\mathbf{P}$ input model for OUT_NAME in $\mathbf{W}$. But that minimal model is just $\bigcap \mathbf{F}(\mathbf{P})$. Similarly for $\mathbf{Q}$. Hence

$$\begin{aligned} \text{OUT\_NAME}(\mathbf{P}) &= \text{OUT\_NAME}^{\cap \mathbf{F}(\mathbf{P})} \\ &\subseteq \text{OUT\_NAME}^{\cap \mathbf{F}(\mathbf{Q})} \\ &= \text{OUT\_NAME}(\mathbf{Q}). \end{aligned}$$

We have been working since Chapter 1 with intersections of families of models. It is also possible to define unions.

Definition. Let $\{\mathbf{I}_1, \mathbf{I}_2, \mathbf{I}_3, \ldots\}$ be a family of interpretations. By the *union* of this family, written $\bigcup\{\mathbf{I}_1, \mathbf{I}_2, \mathbf{I}_3, \ldots\}$, we mean the interpretation $\mathbf{J}$ that assigns to each identifier the union of the relations that $\mathbf{I}_1$, $\mathbf{I}_2$, $\mathbf{I}_3, \ldots$ assign to it.

For example, consider again the interpretations $\mathbf{I}_0$ to $\mathbf{I}_4$ defined in Chapter 1, §1.7. You should verify that $\bigcup\{\mathbf{I}_1, \mathbf{I}_2\} = \mathbf{I}_2$, and $\bigcup\{\mathbf{I}_2, \mathbf{I}_3\} \le \mathbf{I}_4$.

Exercise 3.9.6. Let $\mathbf{F}$ be a family of interpretations, and suppose $\mathbf{I} \in \mathbf{F}$. Show $\mathbf{I} \le \bigcup\mathbf{F}$.

While the intersection of a family of models must be a model, the same is not always true of unions. For example, suppose we are working with the language EFS(**integer**) and we have the following simple procedure.

B(1):
    A(0)→A(1)→B(2).

Let $\mathbf{I}$ and $\mathbf{J}$ be interpretations that assign to SUC the successor relation (so both are interpretations in the basic work space of EFS(**integer**)). Also let

$$A^{\mathbf{I}} = \{0\}$$
$$A^{\mathbf{J}} = \{1\}$$
$$B^{\mathbf{I}} = B^{\mathbf{J}} = \varnothing$$

and on other identifiers $\mathbf{I}$ and $\mathbf{J}$ are arbitrary. Both $\mathbf{I}$ and $\mathbf{J}$ are *models* for B. $\mathbf{I}$ is, for example, because A(1) is *not* true under $\mathbf{I}$, hence A(0)→A(1)→B(2) *is* true. But if $\mathbf{K} = \bigcup\{\mathbf{I}, \mathbf{J}\}$ then

$$A^{\mathbf{K}} = A^{\mathbf{I}} \cup A^{\mathbf{J}} = \{0, 1\}$$
$$B^{\mathbf{K}} = B^{\mathbf{I}} \cup B^{\mathbf{J}} = \varnothing$$

and so A(0) and A(1) are both true under $\mathbf{K}$ while B(2) is not. Thus, $\mathbf{K}$ is *not* a model for B.

Definition. Let $\mathbf{F} = \{\mathbf{I}_1, \mathbf{I}_2, \mathbf{I}_3, \ldots\}$ be a family of interpretations. It is called a *directed* family if, for any $\mathbf{I}_i$, $\mathbf{I}_j \in \mathbf{F}$, there is some $\mathbf{I}_k \in \mathbf{F}$ with $\mathbf{I}_i \le \mathbf{I}_k$ and $\mathbf{I}_j \le \mathbf{I}_k$.

For example, the family of *all* interpretations is directed. For any two interpretations $\mathbf{I}_i$ and $\mathbf{I}_j$, just take $\mathbf{I}_k = \bigcup\{\mathbf{I}_i, \mathbf{I}_j\}$. The family $\mathbf{I}_0$ to $\mathbf{I}_4$ from Chapter 1, §1.7 is directed. We will provide another example in the proof of Proposition 3.9.3.

Proposition 3.9.2. *Let* $\mathbf{F}$ *be a family of* ($\mathbf{P}$ *input*) *models for a procedure OUT_NAME in work space* $\mathbf{W}$. *If* $\mathbf{F}$ *is directed, then* $\bigcup\mathbf{F}$ *is also a* ($\mathbf{P}$ *input*) *model for OUT_NAME in* $\mathbf{W}$.

Proof. Suppose **F** is directed. We only check the condition that each substitution instance of a procedure statement of OUT_NAME should be true under $\bigcup\mathbf{F}$. Let $X_1 \to X_2 \to \ldots \to X_p \to T$ be a substitution instance of a procedure statement of OUT_NAME (where each X_i and T is atomic). Suppose each of $X_1, X_2, \ldots, X_p$ is true under $\bigcup\mathbf{F}$. We show T is true under $\bigcup\mathbf{F}$.

Since X_1 is true under $\bigcup\mathbf{F}$, there must be some interpretation in the family **F** that makes X_1 true, say $\mathbf{I}_1$. Similarly, there must be interpretations $\mathbf{I}_2, \ldots, \mathbf{I}_p$ in the family making each of $X_2, \ldots, X_p$ respectively true. The following diagram displays the arrangement described in the next part of the proof.

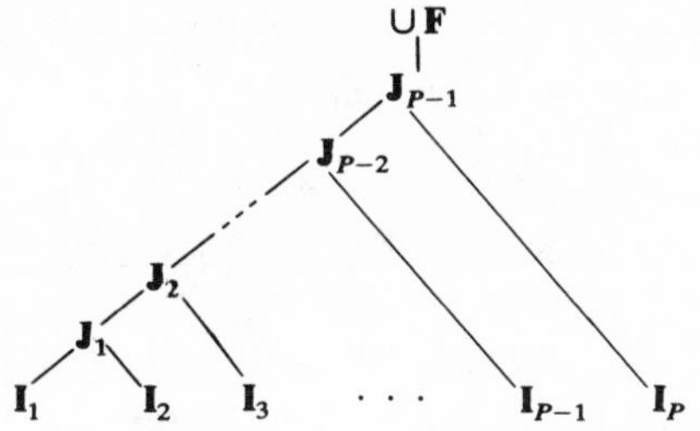

Since $\mathbf{I}_1$ and $\mathbf{I}_2$ are in the family **F**, which is directed, for some $\mathbf{J}_1$ in the family, $\mathbf{I}_1 \le \mathbf{J}_1$ and $\mathbf{I}_2 \le \mathbf{J}_1$. Then $\mathbf{J}_1$ must make both X_1 and X_2 true, by Exercise 3.9.2. Next, since $\mathbf{J}_1$ and $\mathbf{I}_3$ are in the family, for some $\mathbf{J}_2$ in the family, $\mathbf{J}_1 \le \mathbf{J}_2$ and $\mathbf{I}_3 \le \mathbf{J}_2$ (and hence $\mathbf{I}_1 \le \mathbf{J}_2$ and $\mathbf{I}_2 \le \mathbf{J}_2$ by transitivity). And $\mathbf{J}_2$ must make all of X_1, X_2, X_3 true. Continuing in this way, there must a model $\mathbf{J}_{p-1}$ in the family with all of $\mathbf{I}_1 \le \mathbf{J}_{p-1}, \ldots, I_p \le \mathbf{J}_{p-1}$, and such that $\mathbf{J}_{p-1}$ makes all of $X_1, X_2, \ldots, X_p$ true. Since $\mathbf{J}_{p-1}$ is in the family **F**, and members of the family are *models* for OUT_NAME, $X_1 \to X_2 \to \ldots \to X_p \to T$ is true under $\mathbf{J}_{p-1}$. It follows that T is true under $\mathbf{J}_{p-1}$. But then, since $\mathbf{J}_{p-1} \in \mathbf{F}$, $\bigcup\mathbf{F}$ makes T true also, using Exercises 3.9.2 and 3.9.6.

Exercise 3.9.7. For each **R**, let **M(R)** be the minimal **R** input model for the procedure

OUT_NAME (n) INPUT IN_NAME (k):
(procedure body of OUT_NAME).

in work space **W**. Show that, if $\mathbf{P} \subseteq \mathbf{Q}$ then $\mathbf{M(P)} \le \mathbf{M(Q)}$. *Hint*: Use Exercise 3.9.4.

Proposition 3.9.3. *An operator defined by a procedure is compact.* (*This is Proposition* 3.7.2 *again.*)

Proof. Once again, consider the following procedure in work space **W**.

OUT_NAME (n) INPUT IN_NAME (n):
(procedure body of OUT_NAME).

For each n-place relation $\mathbf{R}$ on the domain of $\mathbf{W}$, let $\mathbf{M(R)}$ be the minimal $\mathbf{R}$ input model for OUT_NAME in $\mathbf{W}$. By Corollary 3.8.3, for each $\mathbf{R}$, OUT_NAME($\mathbf{R}$) = OUT_NAME$^{\mathbf{M(R)}}$. Now, let $\mathbf{P}$ be an n-place relation on the domain of $\mathbf{W}$, fixed for the rest of this proof. We will show that if $\langle t_1, \ldots, t_n \rangle \in$ OUT_NAME($\mathbf{P}$), then for some *finite* $\mathbf{Q} \subseteq \mathbf{P}$, $\langle t_1, \ldots, t_n \rangle \in$ OUT_NAME($\mathbf{Q}$). Equivalently, we will show that if $\langle t_1, \ldots, t_n \rangle \in$ OUT_NAME$^{\mathbf{M(P)}}$, then for some finite $\mathbf{Q} \subseteq \mathbf{P}$, $\langle t_1, \ldots, t_n \rangle \in$ OUT_NAME$^{\mathbf{M(Q)}}$.

Let $\mathbf{F} = \{\mathbf{M(Q)} \mid \mathbf{Q} \subseteq \mathbf{P},\ \mathbf{Q}$ finite$\}$. $\mathbf{F}$ is a family of minimal $\mathbf{Q}$ input models, one for each finite subset $\mathbf{Q}$ of $\mathbf{P}$. If $\mathbf{M(Q_1)}$ and $\mathbf{M(Q_2)}$ are two members of $\mathbf{F}$, then $\mathbf{Q}_1$ and $\mathbf{Q}_2$ are finite subsets of $\mathbf{P}$, hence so is $\mathbf{Q}_1 \cup \mathbf{Q}_2$, and $\mathbf{M(Q_1 \cup Q_2)}$ is also a member of $\mathbf{F}$. Since $\mathbf{Q}_1 \subseteq \mathbf{Q}_1 \cup \mathbf{Q}_2$, by Exercise 3.9.7, $\mathbf{M(Q_1)} \le \mathbf{M(Q_1 \cup Q_2)}$. Similarly for $\mathbf{M(Q_2)}$. We have shown that $\mathbf{F}$ is directed. Then Proposition 3.9.2 applies; $\bigcup \mathbf{F}$ is also a model for OUT_NAME in $\mathbf{W}$. By definition, IN_NAME$^{\cup \mathbf{F}}$ is the union of IN_NAME$^{\mathbf{M(Q)}}$ for all $\mathbf{M(Q)} \in \mathbf{F}$. But IN_NAME$^{\mathbf{M(Q)}} = \mathbf{Q}$ (Proposition 3.8.1). So IN_NAME$^{\cup \mathbf{F}} = \bigcup\{\mathbf{Q} \mid \mathbf{Q} \subseteq \mathbf{P},\ \mathbf{Q}$ finite$\}$, that is, IN_NAME$^{\cup \mathbf{F}} = \mathbf{P}$. Then $\bigcup \mathbf{F}$ is a $\mathbf{P}$ input model for OUT_NAME in $\mathbf{W}$.

Now by Exercise 3.9.4, $\mathbf{M(P)} \le \bigcup \mathbf{F}$, so if $\langle t_1, \ldots, t_n \rangle \in$ OUT_NAME$^{\mathbf{M(P)}}$, then $\langle t_1, \ldots, t_n \rangle \in$ OUT_NAME$^{\cup \mathbf{F}}$. By definition of union, this means $\langle t_1, \ldots, t_n \rangle \in$ OUT_NAME$^{\mathbf{M(Q)}}$ for some $\mathbf{M(Q)} \in \mathbf{F}$, hence for some finite $\mathbf{Q} \subseteq \mathbf{P}$, and this ends the proof.

Finally a result that was not established in §3.7, though it could have been.

Proposition 3.9.4. *If the input to an operator defined by a procedure is a generated relation, so is the output.*

Proof. This time we have two procedures to consider in work space $\mathbf{W}$, one of which accepts input, the other of which does not. Say they are

OUT_NAME (n) INPUT IN_NAME (k):
    (procedure body of OUT_NAME).

and

SOURCE (k):
    (procedure body of SOURCE).

Let us say that **SOURCE** is the output of procedure SOURCE. We show that OUT_NAME(**SOURCE**) is a generated relation.

Since the choice of identifiers used in procedures is arbitrary, we can assume that the two procedures OUT_NAME and SOURCE share no identifiers except for those reserved in $\mathbf{W}$. Now consider the following procedure, where TARGET is a previously unused identifier.

TARGET (n):
    (procedure body of OUT_NAME);
    (procedure body of SOURCE);
    SOURCE($x_1, \ldots, x_k$) → IN_NAME($x_1, \ldots, x_k$);
    OUT_NAME($y_1, \ldots, y_n$) → TARGET($y_1, \ldots, y_n$).

We claim that the output of TARGET is exactly OUT_NAME(**SOURCE**). Let **S** be the minimal model for SOURCE in work space **W**, and let **T** be the minimal model for TARGET. Then **SOURCE** = output of procedure SOURCE = $\text{SOURCE}^{\mathbf{S}}$, and output of procedure TARGET = $\text{TARGET}^{\mathbf{T}}$. We show that OUT_NAME($\text{SOURCE}^{\mathbf{S}}$) = $\text{TARGET}^{\mathbf{T}}$.

Because of the procedure statement SOURCE($x_1, \ldots, x_k$) → IN_NAME($x_1, \ldots, x_k$) in TARGET, in *any* model **I** for TARGET, $\text{SOURCE}^{\mathbf{I}} \subseteq \text{IN\_NAME}^{\mathbf{I}}$. Similarly, $\text{OUT\_NAME}^{\mathbf{I}} \subseteq \text{TARGET}^{\mathbf{I}}$. Because every procedure statement of SOURCE is also a procedure statement of TARGET, every model for TARGET is also a model for SOURCE. In particular, **T** is a model for SOURCE, hence $\mathbf{S} \leq \mathbf{T}$ since **S** is the minimal model for SOURCE. Then $\text{SOURCE}^{\mathbf{S}} \subseteq \text{SOURCE}^{\mathbf{T}}$. We have $\text{SOURCE}^{\mathbf{S}} \subseteq \text{SOURCE}^{\mathbf{T}} \subseteq \text{IN\_NAME}^{\mathbf{T}}$. Also, since all the procedure statements of OUT_NAME occur in TARGET, **T** is a model for OUT_NAME as well. We have now shown that **T** is a $\text{SOURCE}^{\mathbf{S}}$ input model for OUT_NAME. It follows by Exercise 3.8.1 that OUT_NAME($\text{SOURCE}^{\mathbf{S}}$) $\subseteq \text{OUT\_NAME}^{\mathbf{T}}$. But $\text{OUT\_NAME}^{\mathbf{T}} \subseteq \text{TARGET}^{\mathbf{T}}$ by remarks earlier. We thus have half the desired conclusion: OUT_NAME($\text{SOURCE}^{\mathbf{S}}$) $\subseteq \text{TARGET}^{\mathbf{T}}$.

For the converse direction, let **M** be the minimal $\text{SOURCE}^{\mathbf{S}}$ input model for OUT_NAME in **W**. Then, of course, OUT_NAME($\text{SOURCE}^{\mathbf{S}}$) = $\text{OUT\_NAME}^{\mathbf{M}}$. The general plan is to extend **M** to a model for the procedure TARGET by combining it with the minimal model for SOURCE.

Define an interpretation **K** to be $\bigcup\{\mathbf{S}, \mathbf{M}\}$. Since SOURCE and OUT_NAME share no identifiers except reserved ones, it follows that, on the unreserved identifiers of SOURCE, **S** and **K** agree (because an identifier that does not occur in OUT_NAME will be assigned the empty relation by **M**). Similarly, on the unreserved identifiers of OUT_NAME, **M** and **K** agree. And on reserved identifiers of **W**, **S**, **M** and **K** all agree. It follows that all procedure statements of (procedure body of OUT_NAME) and of (procedure body of SOURCE) will be true under **K** since they were true under **M** and **S**, respectively. Further, since **M** was a $\text{SOURCE}^{\mathbf{S}}$ input model, $\text{SOURCE}^{\mathbf{S}} \subseteq \text{IN\_NAME}^{\mathbf{M}} = \text{IN\_NAME}^{\mathbf{K}}$. Also $\text{SOURCE}^{\mathbf{S}} = \text{SOURCE}^{\mathbf{K}}$, so $\text{SOURCE}^{\mathbf{K}} \subseteq \text{IN\_NAME}^{\mathbf{K}}$, which means that every substitution instance of SOURCE($x_1, \ldots, x_k$) → IN_NAME($x_1, \ldots, x_k$) is true in **K**.

We do not quite have a model for the procedure TARGET yet. Since the identifier TARGET did not occur in either SOURCE or OUT_NAME, the interpretation of TARGET under **K** is the empty relation. Now define one last interpretation **J** to be like **K** on all identifiers except TARGET, and on that, $\text{TARGET}^{\mathbf{J}}$ is to be $\text{OUT\_NAME}^{\mathbf{K}}$. Then on all procedure statements that do not involve the identifier TARGET, **J** and **K** will behave alike. Hence **J** is also a model for all of (procedure body of OUT_NAME), (procedure body of SOURCE) and SOURCE($x_1, \ldots, x_k$) → IN_NAME($x_1, \ldots, x_k$). And, by definition, **J** is a model for OUT_NAME($y_1, \ldots, y_n$) → TARGET($y_1, \ldots, y_n$) since $\text{TARGET}^{\mathbf{J}} = \text{OUT\_NAME}^{\mathbf{K}} = \text{OUT\_NAME}^{\mathbf{J}}$. Thus, we have a model **J** for the procedure TARGET. By Proposition 1.6.1, the output of the procedure

TARGET, $\text{TARGET}^{\mathbf{T}}$, is a subset of $\text{TARGET}^{\mathbf{J}}$. Then

$$\begin{aligned}\text{TARGET}^{\mathbf{T}} &\subseteq \text{TARGET}^{\mathbf{J}}\\ &= \text{OUT\_NAME}^{\mathbf{K}}\\ &= \text{OUT\_NAME}^{\mathbf{M}}\\ &= \text{OUT\_NAME}(\text{SOURCE}^{\mathbf{S}}).\end{aligned}$$

This concludes the proof.

Exercise 3.9.8. Reestablish the result above, with a proof using augmented traces instead of models.

3.10 THE FIRST RECURSION THEOREM

We have established (twice now) that operators defined by procedures are monotone. Such operators have least fixed points. Also, operators defined by procedures are compact, so we have a method for approximating to least fixed points (outlined in §3.4). We take a closer look at what this method gives us.

Let the following be a procedure in work space **W**, fixed for the rest of this discussion.

OUT_NAME (*n*) INPUT IN_NAME (*n*):
(procedure body of OUT_NAME).

Suppose we construct the approximation sequence for OUT_NAME, according to the rules

$$\begin{aligned}\mathbf{M}_0 &= \varnothing\\ \mathbf{M}_{k+1} &= \text{OUT\_NAME}(\mathbf{M}_k).\end{aligned}$$

Now $\varnothing$ is trivially generated (use a procedure that never assigns anything to the output identifier; a procedure with no procedure statements will do.) Then $\mathbf{M}_0$ is generated. If $\mathbf{M}_k$ is generated so is $\mathbf{M}_{k+1}$ by Proposition 3.9.4. So, by induction, every member of the approximation sequence is a generated relation.

Not only is every member of the approximation sequence generated, but we have a method for writing a procedure for each one. We have just said how to generate $\mathbf{M}_0$. And if we know how to generate $\mathbf{M}_k$ the proof of Proposition 3.9.4 tells us *how* to generate $\text{OUT\_NAME}(\mathbf{M}_k) = \mathbf{M}_{k+1}$. The theorem we are about to prove says the situation just described extends to the limit $\mathbf{M}_\omega$ of the approximation sequence as well. Proposition 3.4.3 says this limit is the least fixed point of the OUT_NAME operator. We show this least fixed point is generated, and we say how to write a procedure for it. This result, due to Kleene, is of fundamental importance. We devote this and the next section to proofs. Once again, proofs based on augmented traces and proofs based on models are both available to us. We supply both for so important a theorem. And this time, for the sake of variety, we begin with a proof that uses models.

The First Recursion Theorem 3.10.1. *The least fixed point of an operator defined by a procedure is a generated relation.*

Proof. Let the following be a procedure in work space **W**:

OUT_NAME (n) INPUT IN_NAME (n):
(procedure body of OUT_NAME).

The discussion above suggests a plan. The approximation sequence is created by repeated applications of OUT_NAME, using the output at one stage as input for the next. We add a procedure statement to OUT_NAME to embody this recycling of output as new input. Let LEAST be an identifier, unused till now. Consider the following procedure.

LEAST (n):
(procedure body of OUT_NAME);
OUT_NAME($x_1, \ldots, x_n$)→IN_NAME($x_1, \ldots, x_n$);
OUT_NAME($x_1, \ldots, x_n$)→LEAST($x_1, \ldots, x_n$).

We show the output of LEAST is exactly the least fixed point of OUT_NAME. The rest of this proof can be looked at as a correctness and completeness argument, where correctness means every member of the output of LEAST is in the least fixed point, and completeness means the output of LEAST includes all the members of the least fixed point. In the following, we use **LEAST** to denote the output of the procedure LEAST, and **FIX** to denote the least fixed point of OUT_NAME. We show **LEAST** = **FIX**.

We begin with a few simple observations. Since OUT_NAME($x_1, \ldots, x_n$)→ IN_NAME($x_1, \ldots, x_n$) is a procedure statement of LEAST, then in any model **I** for LEAST in **W** we must have $\text{OUT\_NAME}^{\mathbf{I}} \subseteq \text{IN\_NAME}^{\mathbf{I}}$. Similarly, because OUT_NAME($x_1, \ldots, x_n$)→LEAST($x_1, \ldots, x_n$) is a procedure statement, we must have $\text{OUT\_NAME}^{\mathbf{I}} \subseteq \text{LEAST}^{\mathbf{I}}$. Further, the *only* procedure statement in which the identifier LEAST occurs is this one. It follows that in the *minimal* model **L** for LEAST, $\text{OUT\_NAME}^{\mathbf{L}} = \text{LEAST}^{\mathbf{L}} = \mathbf{LEAST}$.

Completeness argument: We show **FIX** ⊆ **LEAST.**

Let **L** be the minimal model for LEAST in **W**. Then **L** is a model for all the procedure statements of OUT_NAME and $\mathbf{LEAST} = \text{LEAST}^{\mathbf{L}} = \text{OUT\_NAME}^{\mathbf{L}} \subseteq \text{IN\_NAME}^{\mathbf{L}}$, which means that **L** is a **LEAST** input model for OUT_NAME. Then, by Exercise 3.8.1, OUT_NAME(**LEAST**) ⊆ $\text{OUT\_NAME}^{\mathbf{L}}$, so OUT_NAME(**LEAST**) ⊆ **LEAST**. Thus, **LEAST** is a semifixed point of OUT_NAME, and by Theorem 3.3.3, **FIX** ⊆ **LEAST**.

Correctness argument: We show **LEAST** ⊆ **FIX**.

Let **M(FIX)** be the smallest **FIX** input model for OUT_NAME. Let **M′** be like **M(FIX)** on all identifiers except LEAST, and let $\text{LEAST}^{\mathbf{M}'} = \text{OUT\_NAME}^{\mathbf{M(FIX)}}$. Since **M′** and **M(FIX)** agree on all identifiers except LEAST, and **M(FIX)** is a model for OUT_NAME, which does not contain LEAST, then all procedure statements of (procedure body of OUT_NAME) are true under **M**′. Further, by definition of **M**′, OUT_NAME($x_1, \ldots, x_n$)→LEAST($x_1, \ldots, x_n$) is true under **M**′. Finally,

$$\begin{aligned}\text{OUT\_NAME}^{\mathbf{M(FIX)}} &= \text{OUT\_NAME}(\mathbf{FIX}) \text{ (by Corollary 3.8.3)}\\ &= \mathbf{FIX} \text{ (since } \mathbf{FIX} \text{ is a fixed point)}\\ &= \text{IN\_NAME}^{\mathbf{M(FIX)}} \text{ (by Proposition 3.8.1)}\end{aligned}$$

and this means that **M(FIX)**, and hence **M**′, is a model for OUT_NAME($x_1, \ldots, x_n$)→IN_NAME($x_1, \ldots, x_n$). We have just verified that **M**′ is a model for LEAST. Then, if **L** is the minimal model for LEAST, $\mathbf{L} \le \mathbf{M}'$, and so

$$\begin{aligned}\mathbf{LEAST} &= \text{LEAST}^{\mathbf{L}}\\ &\subseteq \text{LEAST}^{\mathbf{M}'}\\ &= \text{OUT\_NAME}^{\mathbf{M(FIX)}}\\ &= \mathbf{FIX}.\end{aligned}$$

Thus, $\mathbf{LEAST} \subseteq \mathbf{FIX}$. This concludes the proof.

Exercise 3.10.1. Show, using the procedure LEAST constructed in the proof above, that IN_NAME and LEAST all represent the same relation.

Exercise 3.10.2. Let NAME be a procedure in which the two unreserved identifiers ONE and TWO occur, along with possibly other identifiers. Suppose that, using procedure NAME in work space **W**, both ONE and TWO represent the same relation. Create a procedure NEWNAME from NAME by replacing every occurrence of the identifier TWO by an occurrence of ONE. Prove that ONE represents the same relation using either procedure NAME or procedure NEWNAME.

In §3.6 we gave a simple example of an input accepting procedure in the basic work space of EFS(**str**({1})). We repeat it for convenience.

OUT (1) INPUT IN (1):
  OUT(Δ);
  IN(x)→CON(x, 11, y)→OUT(y).

As Exercise 3.7.1 you were asked to determine the least fixed point of OUT. If you did that exercise, you verified that it is the set of even-lengthed words. Now, from the proof of Theorem 3.10.1, we get the following procedure whose output is the least fixed point of OUT.

LEAST (1):
  OUT(Δ);
  IN(x)→CON(x, 11, y)→OUT(y);
  OUT(x)→IN(x);
  OUT(x)→LEAST(x).

Hence the output of LEAST must be the set of even-lengthed words.

By Exercise 3.10.1, all of IN, OUT, and LEAST represent the same set, using the procedure above, hence, by Exercise 3.10.2, we can replace IN and OUT by LEAST without changing the set represented. If we do, we get the following.

LEAST (1):
  LEAST(Δ);
  LEAST(x)$\rightarrow$CON(x, 11, y)$\rightarrow$LEAST(y);
  LEAST(x)$\rightarrow$LEAST(x);
  LEAST(x)$\rightarrow$LEAST(x).

Obviously the last two procedure statements are useless. If we eliminate them we get the following.

LEAST (1):
  LEAST(Δ);
  LEAST(x)$\rightarrow$CON(x, 11, y)$\rightarrow$LEAST(y).

And we are guaranteed that the output of this procedure is the least fixed point of the operator OUT, hence we know the output is the set of even-lengthed words.

Notice that the final version of LEAST we arrived at is, except for choice of identifier, the procedure EVEN from Chapter 1, §1.6. We used it there to illustrate correctness and completeness arguments. What we have just done is given an alternate proof, this time using operators, of the fact that the output is the set of even-lengthed words. This is a technique we will develop further in §3.12.

3.11 A SECOND PROOF

The First Recursion Theorem is important and many kinds of proofs for it have been devised. We gave one using models in §3.10. Now we give a second one using augmented traces as a tool. The underlying idea is still the same, however. One shows a certain procedure actually has the behavior claimed for it.

The First Recursion Theorem (again). *The least fixed point of an operator defined by a procedure is a generated relation.*

Proof. Just as in §3.10, let the following be a procedure in work space **W**.

OUT_NAME (n) INPUT IN_NAME (n):
  (procedure body of OUT_NAME).

Choose an unused identifier LEAST, and write the following procedure.

LEAST (n):
  (procedure body of OUT_NAME);
  OUT_NAME($x_1, \ldots, x_n$)$\rightarrow$IN_NAME($x_1, \ldots, x_n$);
  OUT_NAME($x_1, \ldots, x_n$)$\rightarrow$LEAST($x_1, \ldots, x_n$).

Let **LEAST** be the output of the procedure LEAST, and let **FIX** be the least fixed point of the operator defined by OUT_NAME. We show **LEAST** = **FIX**.

The identifier LEAST occurs in the procedure LEAST only in the procedure statement OUT_NAME($x_1, \ldots, x_n$)→LEAST($x_1, \ldots, x_n$) so if LEAST($t_1, \ldots, t_n$) occurs as a line in a trace for LEAST, OUT_NAME($t_1, \ldots, t_n$) must also occur. Also, if OUT_NAME($t_1, \ldots, t_n$) occurs as a line, OUT_NAME($t_1, \ldots, t_n$)→LEAST($t_1, \ldots, t_n$) and then LEAST($t_1, \ldots, t_n$) could be added. Consequently, the identifiers LEAST and OUT_NAME represent the same thing using the procedure LEAST. It will be more convenient for us to work with the identifier OUT_NAME in the argument below, rather than with LEAST itself. Suppose we have a trace for LEAST, ending in OUT_NAME($t_1, \ldots, t_n$). It is easy to check that if we remove any line in which the identifier LEAST occurs (if any) we still have a correct trace ending in OUT_NAME($t_1, \ldots, t_n$). (This is because the identifier LEAST is not used as a condition in any procedure statement.) Let us call a trace in which LEAST does not occur on any line a *LEAST-free* trace.

So far we have established that, using the procedure LEAST, the identifiers OUT_NAME and LEAST represent the same relations, and further, $\langle t_1, \ldots, t_n \rangle$ is in the relation that OUT_NAME represents if and only if OUT_NAME($t_1, \ldots, t_n$) is the last line of a LEAST-free trace if and only if $\langle t_1, \ldots, t_n \rangle \in$ **LEAST**. To establish the theorem it will be enough to show: $\langle t_1, \ldots, t_n \rangle \in$ **FIX** if and only if OUT_NAME($t_1, \ldots, t_n$) is the last line of a LEAST-free trace for LEAST. The proof divides into two halves, much as it did in §3.10.

Correctness argument: We show, if OUT_NAME($t_1, \ldots, t_n$) is the last line of a LEAST-free trace for LEAST, then $\langle t_1, \ldots, t_n \rangle \in$ **FIX**. This establishes that **LEAST** $\subseteq$ **FIX**. The argument used induction on trace length.

Induction hypothesis: Whenever there is a LEAST-free trace for LEAST ending in OUT_NAME($t_1, \ldots, t_n$) with $< k$ lines, then $\langle t_1, \ldots, t_n \rangle \in$ **FIX**.

To be shown: If we have a LEAST-free trace for LEAST, ending with OUT_NAME($t_1, \ldots, t_n$), with exactly k lines then $\langle t_1, \ldots, t_n \rangle \in$ **FIX**.

Let T be a LEAST-free trace for LEAST, ending with OUT_NAME($t_1, \ldots, t_n$), with exactly k lines. We define two (finite) relations, using the trace T, as follows. Let **IN** be the set of n-tuples $\langle u_1, \ldots, u_n \rangle$ such that IN_NAME($u_1, \ldots, u_n$) occurs as a line of T. And let **OUT** be the set of n-tuples $\langle u_1, \ldots, u_n \rangle$ such that OUT_NAME($u_1, \ldots, u_n$) occurs as a line of T *other than the last line*. We show **IN** $\subseteq$ **OUT** $\subseteq$ **FIX**.

We start by showing **IN** $\subseteq$ **OUT**. Suppose that $\langle u_1, \ldots, u_n \rangle \in$ **IN**. Then IN_NAME($u_1, \ldots, u_n$) occurs as a line in T. Now T is a trace for LEAST, and in the procedure statements of LEAST, IN_NAME occurs in the assignment position of only one: OUT_NAME($x_1, \ldots, x_n$)→IN_NAME($x_1, \ldots, x_n$). (Recall, when the procedure OUT_NAME was written, we were not allowed to use IN_NAME in the assignment position of any of its procedure statements.) Consequently if IN_NAME($u_1, \ldots, u_n$) occurs as a line of T there must be earlier lines OUT_NAME($u_1, \ldots, u_n$)→IN_NAME($u_1, \ldots, u_n$) and OUT_NAME($u_1, \ldots, u_n$), from which we got IN_NAME($u_1, \ldots, u_n$) by the Assignment Rule. But then OUT_NAME($u_1, \ldots, u_n$) must occur in T before the last line, and so $\langle u_1, \ldots, u_n \rangle \in$ **OUT**. We have verified that **IN** ⊆ **OUT**.

Next, we show that **OUT** ⊆ **FIX**. Suppose that $\langle u_1, \ldots, u_n \rangle \in$ **OUT**. Then OUT_NAME($u_1, \ldots, u_n$) occurs as a line of T before the last line. But that means there is a trace with $< k$ lines, ending in OUT_NAME($u_1, \ldots, u_n$). Then, by the induction hypothesis, $\langle u_1, \ldots, u_n \rangle \in$ **FIX**. We have verified that **OUT** ⊆ **FIX**.

T is a trace for LEAST ending with OUT_NAME($t_1, \ldots, t_n$). We now want to convert T into an *augmented* trace for OUT_NAME (for some appropriate input relation) with the same last line. T is LEAST-free, so the procedure statement OUT_NAME($x_1, \ldots, x_n$)→LEAST($x_1, \ldots, x_n$) was not used in T. But T may still not be an augmented trace for OUT_NAME(**P**) (no matter what the input **P**) because OUT_NAME($x_1, \ldots, x_n$)→IN_NAME($x_1, \ldots, x_n$) may have been used, which is a statement of LEAST but not of OUT_NAME. Well, let T' be like T but with all substitution instances of OUT_NAME($x_1, \ldots, x_n$)→IN_NAME($x_1, \ldots, x_n$) removed. Now T' still may not be a correct trace for OUT_NAME because, in T, we may have *used* instances of OUT_NAME($x_1, \ldots, x_n$)→IN_NAME($x_1, \ldots, x_n$) and the Assignment Rule to get lines of the form IN_NAME($u_1, \ldots, u_n$). And now we cannot justify the presence of such lines if we are trying for a trace for OUT_NAME. However, if we were *told* that IN_NAME represented **IN**, then we would be allowed such lines in a trace, as instances of a given relation. That is, T' is a correct augmented trace for OUT_NAME(**IN**).

Since the last line of T was OUT_NAME($t_1, \ldots, t_n$), that is still the last line of T'. We have just shown that $\langle t_1, \ldots, t_n \rangle \in$ OUT_NAME(**IN**). But we have also shown **IN** ⊆ **FIX**, so by monotonicity, OUT_NAME(**IN**) ⊆ OUT_NAME(**FIX**), and hence $\langle t_1, \ldots, t_n \rangle \in$ OUT_NAME(**FIX**). But **FIX** is a fixed point, so $\langle t_1, \ldots, t_n \rangle \in$ **FIX**. This is the item we wished to show; the induction step is complete. It now follows by ordinary induction that if OUT_NAME($t_1, \ldots, t_n$) is the last line of a LEAST-free trace for LEAST (of any length) then $\langle t_1, \ldots, t_n \rangle \in$ **FIX**, and this implies that **LEAST** ⊆ **FIX**.

Completeness argument: We show that **FIX** ⊆ **LEAST**, concluding the proof. By the Generalized Induction Theorem (3.3.3), it is enough to show that OUT_NAME(**LEAST**) ⊆ **LEAST**.

Suppose that $\langle t_1, \ldots, t_n \rangle \in$ OUT_NAME(**LEAST**). Then there must be an augmented trace of OUT_NAME(**LEAST**) ending in OUT_NAME($t_1, \ldots, t_n$); call the trace A. The only thing that prevents A from being a trace for LEAST is that there may be lines IN_NAME($u_1, \ldots, u_n$) where $\langle u_1, \ldots, u_n \rangle \in$ **LEAST**, whose presence would not be justified in a trace for LEAST. What we do is rewrite A to turn it into a correct trace for LEAST. Consider one such "problem line" in A, say IN_NAME($u_1, \ldots, u_n$) where $\langle u_1, \ldots, u_n \rangle \in$ **LEAST**. Since

$\langle u_1, \ldots, u_n \rangle \in \mathbf{LEAST}$, there is a LEAST-free trace for LEAST ending in OUT_NAME($u_1, \ldots, u_n$); let us call this trace B. It, of course, ends with OUT_NAME($u_1, \ldots, u_n$). We add two more lines, the first a substitution instance of a procedure statement of LEAST, the second coming from earlier lines by the Assignment Rule.

{trace B}
OUT_NAME($u_1, \ldots, u_n$)
OUT_NAME($u_1, \ldots, u_n$)$\rightarrow$IN_NAME($u_1, \ldots, u_n$)
IN_NAME($u_1, \ldots, u_n$).

This is a correct, LEAST-free trace for LEAST. In the trace A, where the line IN_NAME($u_1, \ldots, u_n$) occurred, replace it with the lines above. Do a similar thing with every line of A of the form IN_NAME($v_1, \ldots, v_n$). Call the result A'. It is not hard to see that A' will be a correct LEAST-free trace for LEAST, with no "unjustified" inputs from **LEAST** left. Since it still ends with OUT_NAME($t_1, \ldots, t_n$), we have that $\langle t_1, \ldots, t_n \rangle \in \mathbf{LEAST}$. Thus, OUT_NAME(**LEAST**) $\subseteq$ **LEAST** and we are done.

3.12 APPLICATIONS

What the First Recursion Theorem says is this. If we want the smallest relation **S** that meets certain closure conditions, and if those conditions can be stated in EFS language (in the form of an input accepting procedure), then a procedure that generates **S** can be written. This is often a natural way of arriving at a procedure for **S**, and verifying its behavior at the same time. In Chapter 5, §5.5 and in Chapter 6, §6.11 are examples of applications of the First Recursion Theorem that illustrate this. Things are especially simple if **S** is the graph of a function. In this section, we demonstrate the technique with an easy example, the factorial function in arithmetic. We begin with some general terminology.

Definition. Let **R** be an $n+1$ place relation. **R** is the *graph* of a function provided, whenever $\langle x_1, \ldots, x_n, y \rangle \in \mathbf{R}$ and $\langle x_1, \ldots, x_n, z \rangle \in \mathbf{R}$ then $y = z$.

In other words, for **R** to be the graph of a function, **R** must be single valued in the sense that, if there are any y's such that $\langle x_1, \ldots, x_n, y \rangle \in \mathbf{R}$, then there is only one such y.

Definition. Suppose that **R** is the graph of a function. The *domain* of **R** is the set of n-tuples $\langle x_1, \ldots, x_n \rangle$ such that, for some y, $\langle x_1, \ldots, x_n, y \rangle \in \mathbf{R}$. And **R** is the *graph of* that n-place function f whose domain is that of **R**, and is such that, for $\langle x_1, \ldots, x_n \rangle$ in the domain, $f(x_1, \ldots, x_n)$ is that unique y for which $\langle x_1, \ldots, x_n, y \rangle \in \mathbf{R}$.

For example, let **R** be the collection of triples of nonnegative integers $\langle x, y, z \rangle$ such that $x \geq y$, and $z = x - y$. **R** is the graph of a function. The domain of **R** is the set of all $\langle x, y\rangle$ with $x \geq y$. The function whose graph **R** is, is usually called the subtraction function. The domain of the graph of a function need not include every n-tuple. In the subtraction example, $\langle 3, 4\rangle$ is not in the domain. Division is another such example, since one cannot divide by 0.

Definition. Let **R** be an $n + 1$ place relation on **D**, and suppose **R** is the graph of a function f. If every n-tuple from **D** is in the domain of **R**, the function f is called a *total* function. A function that is not total is called *partial.*

Proposition 3.12.1. *Let Φ be a monotone operator. Suppose some function graph **R** is a semifixed point of Φ, that is, $\Phi(\mathbf{R}) \subseteq \mathbf{R}$. Then the least fixed point of Φ is a function graph.*

Proof. Let **S** be the least fixed point of Φ. By Theorem 3.3.3, $\mathbf{S} \subseteq \mathbf{R}$. Then, if $\langle x_1, \ldots, x_n, y\rangle$ and $\langle x_1, \ldots, x_n, z\rangle$ are in **S**, they are also in **R**, and since **R** is a function graph, $y = z$.

Exercise 3.12.1. Suppose that **R** and **S** are graphs of functions, $\mathbf{S} \subseteq \mathbf{R}$, and **R** and **S** have the same domains. Show that $\mathbf{R} = \mathbf{S}$.

Enough preliminaries. Now we take up the promised example of the factorial function. We work with EFS(**integer**), in the work space reached by the end of Chapter 2, §2.2. The factorial function can be characterized by the following conditions.

$$0! = 1$$
$$(n + 1)! = (n + 1) \cdot n!$$

The second clause specifies the value of a factorial in terms of a (smaller) factorial. Characterizations like this are often called *recursive.* The key feature is that, instead of an explicit definition, we have a reduction of a problem (evaluate $n + 1!$) to a simpler problem (evaluate $n!$) *of the same kind.* After enough such reductions we get down to $0!$, and the first clause of the characterization takes over.

Now, let **F** be the *graph* of the factorial function. So $\langle x, y\rangle \in \mathbf{F}$ just when $y = x!$. The conditions above give us the following *closure conditions* for **F**.

$$\langle 0, 1\rangle \in \mathbf{F}$$
$$\langle x, y\rangle \in \mathbf{F} \Rightarrow \langle x + 1, (x + 1) \cdot y\rangle \in \mathbf{F}.$$

There are many relations meeting these closure conditions. For example, the relation consisting of *all* ordered pairs of nonnegative integers trivially does. What we show is that the graph of the factorial function is the *smallest* relation meeting the closure conditions. And naturally, we bring operators into the picture.

Let Φ be the operator given by $\Phi(\mathbf{R}) = \{\langle 0, 1\rangle\} \cup \{\langle x + 1, (x + 1) \cdot y\rangle \mid \langle x, y\rangle \in \mathbf{R}\}$.

Exercise 3.12.2. Show the following.

(a) Φ is monotone.
(b) $\mathbf{R}$ is a semifixed point of Φ if and only if $\mathbf{R}$ satisfies the closure conditions

$$\langle 0, 1\rangle \in \mathbf{R}$$
$$\langle x, y\rangle \in \mathbf{R} \Rightarrow \langle x+1, (x+1)\cdot y\rangle \in \mathbf{R}.$$

By part (b) of this exercise, the graph $\mathbf{F}$ of the factorial function is a semifixed point of Φ. And by part (a), Φ has a least fixed point. Then, by Proposition 3.12.1, the least fixed point of Φ must be a function graph.

Exercise. 3.12.3. Show the least fixed point of Φ is the graph of a *total* function. *Hint:* Use induction to show that every nonnegative integer is in the domain.

If $\mathbf{L}$ is the least fixed point of Φ, $\mathbf{L} \subseteq \mathbf{F}$ (since $\mathbf{F}$ is a semifixed point). But both $\mathbf{L}$ and $\mathbf{F}$ have the same domains, so $\mathbf{L} = \mathbf{F}$ by Exercise 3.12.1. That is, the least fixed point of Φ is exactly the graph of the factorial function. Now all we have to do is write a procedure for Φ. Well, just formalize the definition of Φ.

```
OUT_FACT (2) INPUT IN_FACT (2):
    OUT_FACT(0, 1);
    IN_FACT(x, y)→SUC(x, z)→TIMES(z, y, w)→OUT_FACT(z, w).
```

An appeal to the First Recursion Theorem now gives us a procedure to generate the graph of the factorial function. And, not so incidentally, we have already established the correctness and completeness of the procedure we get.

Exercise 3.12.4. Write out the procedure for the graph of the factorial function that the First Recursion Theorem proof supplies, and simplify it using Exercises 3.10.1 and 3.10.2.

Exercise 3.12.5. Consider EFS(**str**($\{a, b, 1\}$)). Let f be the function with domain all strings made up of a's and b's, and such that $f(w)$ is a string of 1's containing as many 1's as w has a's. Let Φ be the operator

$$\Phi(\mathbf{R}) = \{\langle \Delta, \Delta\rangle\} \cup \{\langle xa, y1\rangle \mid \langle x, y\rangle \in \mathbf{R}\} \cup \{\langle xb, y\rangle \mid \langle x, y\rangle \in \mathbf{R}\}.$$

(a) Prove the least fixed point of Φ is the graph of f.
(b) Write a procedure for Φ.
(c) Use the First Recursion Theorem and produce a procedure whose output is the graph of f.

Exercise 3.12.6. Consider EFS(**str**($\{a, b\}$)). Write a procedure defining an operator whose least fixed point is the graph of a function g where $g(w)$ is the word w reversed.

3.13 A NORMAL FORM THEOREM

The converse of the First Recursion Theorem is both true and trivial: Every generated relation is the least fixed point of some input accepting procedure. The proof goes as follows. Suppose we have a generated relation **R**. Say **R** is the output of the following procedure in work space **W**.

RRR (n):
    (procedure body of RRR).

Choose some identifier, III say, not reserved and not used in procedure RRR, and consider the following input accepting procedure, with the same procedure body as above.

RRR (n) INPUT III (n):
    (procedure body of RRR).

In this, III represents input, but is not actually used. So, no matter what the input, the output is the same as that of the earlier RRR procedure that did not accept input, namely **R**. Then **R** is a fixed point of this procedure, the only one, hence the least one.

This is not a very interesting result since the input accepting procedure we got was essentially just the procedure we began with. In particular, any recursive procedure calls present in the original RRR procedure are still present in the input-accepting version. But, it is possible to restrict ourselves to operators of an especially simple form. We will show that all recursive procedure calls can be concentrated into a single application of the First Recursion Theorem, and hence we can get all generated relations by considering operators defined by procedures that allow no "looping back." This is sometimes useful because it makes correctness and completeness arguments easier.

Definition. We say a procedure (that accepts input or not) is *simple* if no identifier occurs in the assignment position of a procedure statement and also as a condition of a procedure statement.

Definition. Let **R** be an $n + k$ place relation on a domain **D**, and suppose that $c_1, \ldots, c_n$ are members of **D**. Let **S** be the k-place relation given by

$$\langle x_1, \ldots, x_k \rangle \in \mathbf{S} \Leftrightarrow \langle c_1, \ldots, c_n, x_1, \ldots, x_k \rangle \in \mathbf{R}.$$

S is called the *section* of **R** at $\langle c_1, \ldots, c_n \rangle$.

We allow n to be 0, that is, we consider a relation to be a section of itself. The notion of section is a special case of what we called "inserting constants" in Chapter 2, §2.6, where it was shown that it turned generated relations into generated relations. The term "section" comes from geometry, as the following illustrates.

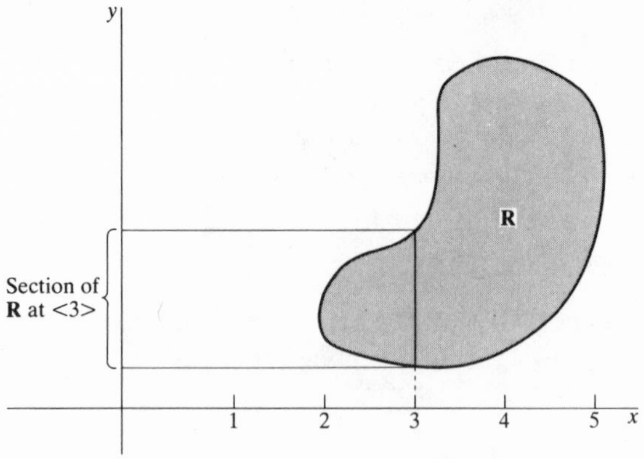

Figure 3.1

For example, suppose that **R** is the addition relation on the nonnegative integers: $\langle x, y, z \rangle \in \mathbf{R}$ if $x + y = z$. Let **S** be the section of **R** at 2; then **S** is the "counting by 2" relation.

Normal Form Theorem 3.13.1. *Every generated relation is a section of the least fixed point of a simple input accepting procedure.*

Proof. The proof, like those earlier, is constructive: Given a procedure with output **S**, we show how to write a simple input accepting procedure whose least fixed point **R** has **S** as a section. Rather than presenting the method in general terms, we work through a representative specific example. Suppose we have work space **W**, s is some member of the domain of **W**, RES_ONE and RES_TWO are reserved identifiers, and the following procedure in the work space is under consideration. (Don't try to figure out the behavior of this procedure; it's merely an arbitrary example.)

TWO (2):
RES_ONE(x, y)→ONE(x);
ONE(x)→RES_TWO(x, s, z)→THREE(x, y, z);
ONE(x)→THREE(x, y, z)→TWO(y, z).

Apart from reserved identifiers, the identifiers are ONE, TWO, and THREE. ONE occurs in the assignment position of the first procedure statement and as a condition in the other two, so the procedure is not simple. The maximum number of places for all unreserved identifiers in this procedure is 3. Choose a new identifier, NEW, which will be used to represent a 4-place relation (one more place than the maximum). Rewrite procedure TWO so that all uses of ONE, TWO, and THREE are replaced by occurrences of NEW as follows. First, choose three distinct members of the work space domain, say a, b, and c. Replace

ONE(x)	by	NEW(a, a, a, x)
TWO(x, y)	by	NEW(b, b, x, y)
THREE(x, y, z)	by	NEW(c, x, y, z).

The first item in the 4-tuple following the identifier NEW lets us keep track of which of the original identifiers we are replacing, by "tagging" with a, b, or c. Since ONE, TWO, and THREE represent relations of different lengths, we have "padded" NEW when necessary, to get a 4-place relation in all cases. Making this replacement, we get the following procedure.

```
NEW (4):
    RES_ONE(x, y) → NEW(a, a, a, x);
    NEW(a, a, a, x) → RES_TWO(x, s, z) → NEW(c, x, y, z);
    NEW(a, a, a, x) → NEW(c, x, y, z) → NEW(b, b, y, z).
```

$\langle b, b, y, z \rangle$ is an output of NEW if and only if $\langle y, z \rangle$ is an output of TWO. More generally, there is a trace for NEW ending in NEW(a, a, a, x) if and only if there is a trace for TWO ending in ONE(x); and similarly for NEW(b, b, x, y) and TWO(x, y); and for NEW(c, x, y, z) and THREE(x, y, z). This is straightforward since the translation we applied to the procedure statements of TWO applies equally well to every line of a trace. Thus, the output of TWO is the section at $\langle b, b \rangle$ of the output of NEW. What we have shown so far is that, in addition to the reserved identifiers, only one additional identifier is needed, provided we allow the taking of sections.

The identifier NEW occurs both in assignment positions and as conditions of procedure statements; NEW is not simple. The next step is to separate the condition and assignment roles of NEW. Let IN and OUT be two unused identifiers. In NEW, wherever the identifier NEW occurs in an assignment position, replace it by OUT; wherever NEW occurs as a condition, replace it by IN. We get a new procedure body, which we supply with a new header, as follows.

```
OUT (4) INPUT IN (4):
    RES_ONE(x, y) → OUT(a, a, a, x);
    IN(a, a, a, x) → RES_TWO(x, s, z) → OUT(c, x, y, z);
    IN(a, a, a, x) → IN(c, x, y, z) → OUT(b, b, y, z).
```

Of course, by construction, OUT is a simple procedure. According to the proof of the First Recursion Theorem, the least fixed point of the operator defined by OUT is given by the following procedure.

```
LEAST (4):
    RES_ONE(x, y) → OUT(a, a, a, x);
    IN(a, a, a, x) → RES_TWO(x, s, z) → OUT(c, x, y, z);
    IN(a, a, a, x) → IN(c, x, y, z) → OUT(b, b, y, z);
```

OUT$(x, y, z, w)\rightarrow$IN(x, y, z, w);
OUT$(x, y, z, w)\rightarrow$LEAST(x, y, z, w).

By Exercise 3.10.1, all of LEAST, IN, and OUT represent the same relation using this procedure. So by Exercise 3.10.2, IN and OUT can be replaced by LEAST without changing the relation that LEAST represents. Doing so turns the last two procedure statements into trivialities, which can be dropped. We are left with the following.

LEAST (4):
RES_ONE$(x, y)\rightarrow$LEAST(a, a, a, x);
LEAST$(a, a, a, x)\rightarrow$RES_TWO$(x, s, z)\rightarrow$LEAST(c, x, y, z);
LEAST$(a, a, a, x)\rightarrow$LEAST$(c, x, y, z)\rightarrow$LEAST(b, b, y, z).

The output of this is the least fixed point of the operator defined by OUT. But the procedure LEAST is the same as the procedure NEW, except for the trivial fact that the identifier LEAST has been used everywhere instead of NEW.

This finishes the proof. We have an operator defined by the simple procedure OUT, with a least fixed point having the output of TWO as a section.

The argument just given shows *some* appropriate input accepting procedure always exists. The method we gave does not always produce the most "natural" such procedure, whatever that means.

Exercise 3.13.1. Consider the procedure SUM from Chapter 2, §2.2. Follow the method above and write a simple procedure for an operator whose least fixed point has the output of SUM as a section. Also see if there is some other input accepting operator that has this behavior, and which seems "nicer."

Exercise 3.13.2. Let **TWO** be the output of procedure TWO above, and **NEW** be the output of NEW. Show *using models* that $\langle x, y\rangle \in$ **TWO** $\Leftrightarrow \langle b, b, x, y\rangle \in$ **NEW**.

3.14 BACKGROUND

For the language EFS(**integer**), the family of operators defined by procedures is well known to mathematicians working in recursion theory. These operators are called *enumeration operators*. The standard definition can be found in [Rogers 1967], along with a general development of their properties. It probably will not be obvious that enumeration operators as defined by Rogers are what we have been talking about here. A proof of equivalence can be found in [Fitting 1981], as part of a more general investigation.

The fixed point theorem for monotone operators is often known as the Knaster-Tarski Theorem. It was first proved in [Knaster 1928] and generalized in [Tarski 1955]. A detailed history of this theorem and extensions of it appears in [Lassez, Nguyen, and Sonenberg 1982].

The First Recursion Theorem is due to Kleene, and may be found in [Kleene 1952], formulated for recursive functionals. In [Rogers 1967] it is established for enumeration operators, though by a different proof than we used. The proof given here has its origins in [Fitting 1981]. There is a remarkable and elegant proof, due to D. Park, in [Scott 1976].

The terminology of *compact* operators is not standard. More commonly, one sees reference to *continuity*. To make clear why this term is used, topological notions must be introduced, and we did not wish to do this here. [Scott 1976] is a fundamental reference for this approach, though the topology used traces back to exercise 11–35 in [Rogers 1967].

REFERENCES

[Fitting 1981] *Fundamentals of Generalized Recursion Theory,* M. Fitting, North Holland Publishing Co., Amsterdam, 1981.

[Kleene 1952] *Introduction To Metamathematics,* S. Kleene, D. Van Nostrand Co., Princeton, N.J., 1952.

[Knaster 1928] Un théorème sur les fonctions d'ensembles, B. Knaster, *Annales de la Société Polonaise de Mathématique,* Vol. 6, pp. 133–134, 1928.

[Lassez, Nguyen, and Sonenberg 1982] Fixed point theorems and semantics: A folk tale, J. Lassez, V. Nguyen, and E. Sonenberg, *Information Processing Letters,* Vol. 14, pp. 112–116, 1982.

[Rogers 1967] *Theory of Recursive Functions and Effective Computability,* H. Rogers, McGraw-Hill, New York, 1967.

[Scott 1976] Data types as lattices, D. Scott, *SIAM Journal of Computing,* Vol. 5, pp. 522–587, 1976.

[Tarski 1955] A lattice-theoretical fixpoint theorem and its applications, A. Tarski, *Pacific Journal of Mathematics,* Vol. 5, pp. 285–309, 1955.

4
IMPLEMENTING DATA STRUCTURES

4.1 INTRODUCTION

Each data structure has its EFS language directly, but we have also seen some indirect data structure use. For example, in Chapter 2, §2.3, when working with EFS(**tree**(**A**)), number representatives were defined and there was an exercise to show the "successor" relation on them was generated. In effect the structure **integer** was being simulated. The same thing happened in Chapter 2, §2.4 with EFS(**set**(**A**)). More generally, suppose we are working with EFS($\langle \mathbf{E}; \mathbf{S}_1, \ldots, \mathbf{S}_v \rangle$), and there is some data structure $\langle \mathbf{D}; \mathbf{R}_1, \ldots, \mathbf{R}_u \rangle$ we wish to consider also. It may be possible to find representatives for the members of **D** among the objects of **E**. Call these *codes* for members of **D**. We may be able to work with the structure $\langle \mathbf{D}; \mathbf{R}_1, \ldots, \mathbf{R}_u \rangle$ while staying within the EFS language for $\langle \mathbf{E}; \mathbf{S}_1, \ldots, \mathbf{S}_v \rangle$, by working with these codes for the members of **D** instead of the "real thing." If this happens, we say we have an *implementation* of $\langle \mathbf{D}; \mathbf{R}_1, \ldots, \mathbf{R}_u \rangle$ in $\langle \mathbf{E}; \mathbf{S}_1, \ldots, \mathbf{S}_v \rangle$.

Suppose we have an implementation of $\langle \mathbf{D}; \mathbf{R}_1, \ldots, \mathbf{R}_u \rangle$ in $\langle \mathbf{E}; \mathbf{S}_1, \ldots, \mathbf{S}_v \rangle$. Does working in EFS($\langle \mathbf{E}; \mathbf{S}_1, \ldots, \mathbf{S}_v \rangle$) with codes give us any more power than we would have if we worked directly with EFS($\langle \mathbf{D}; \mathbf{R}_1, \ldots, \mathbf{R}_u \rangle$)? If the answer is no, we say we have a *conservative* implementation of $\langle \mathbf{D}; \mathbf{R}_1, \ldots, \mathbf{R}_u \rangle$ in $\langle \mathbf{E}; \mathbf{S}_1, \ldots, \mathbf{S}_v \rangle$. With a conservative implementation it does not matter whether we work with original objects or with codes.

Progamming can be looked at as a two-sided activity. First, what relations are generated; what are the possible outputs for procedures? This is what we have been discussing up till now. Second, what other data structures have implementations in the one we have taken as basic? This is the issue now. We will see that most of the data structures considered in earlier sections are much richer than first appearances might suggest. **str**(L), **integer**, **tree**(**A**) and **set**(**A**) each allows the conservative implementation of all these other data structures (under reasonable assumptions about **A**). Then for the rest of this book we can concentrate on one EFS language instead of many.

4.2 CODINGS

Definition. Suppose we have two domains **D** and **E**. We have a *coding* of **D** in **E** if, to each member d of **D**, one or more members of **E** have been assigned, called *codes* for d, so that no two members of **D** share a code. If T is a coding of **D** in **E**, for each $d \in \mathbf{D}$, d^T is the set of codes of d. Each member of d^T is a *code*, or a T-*code* for d. The requirement that no two members of **D** share a code can be stated: if $c \neq d$ then c^T and d^T are disjoint, $c^T \cap d^T = \varnothing$. If T is a coding that assigns just one code to each item, in the interests of simple notation we write $t = d^T$ rather than $t \in d^T$ to indicate that t is *the* T-code of d.

For example, in Chapter 2, §2.3 when working with EFS(**tree**(**A**)) we called certain trees *number representatives*. These number representatives can be thought of as *codes* for the nonnegative integers. A similar example is in Chapter 2, §2.4, again coding the nonnegative integers, but this time as sets. Another, less obvious, example also came up in Chapter 2, §2.4. When we wanted to talk about sets, we used *names* for them. We can think of this as a coding of sets by words. In this example, codes are not unique since sets have many names.

Definition. Suppose we have a coding T of **D** in **E**. If $\langle d_1, \ldots, d_n \rangle$ is an n-tuple of members of **D**, we call $\langle e_1, \ldots, e_n \rangle$ a T-*code* for $\langle d_1, \ldots, d_n \rangle$ provided e_1 is a T-code for $d_1, \ldots, e_n$ is a T-code for d_n. If **R** is an n-place relation on **D**, by $\mathbf{R}^T$ we mean the n-place relation on **E** consisting of all T-codes for members of **R**.

For example, suppose that T is the coding of the nonnegative integers in the domain of **tree**(**A**), mentioned earlier. Let **SUC** be the successor relation on nonnegative integers. Then $\mathbf{SUC}^T$ is the relation on trees that you were asked to write a procedure for as Exercise 2.3.3. The definition of T-code for a relation can be restated as $\langle e_1, \ldots, e_n \rangle \in \mathbf{R}^T$ if there are $d_1, \ldots, d_n \in \mathbf{D}$ such that $\langle d_1, \ldots, d_n \rangle \in \mathbf{R}$ and $e_1 \in (d_1)^T, \ldots, e_n \in (d_n)^T$. For $d \in \mathbf{D}$, $\{d\}$ is a 1-place relation on **D**, and we have defined things so that if T is a coding of **D** in **E**, then $d^T = \{d\}^T$.

Exercise 4.2.1. Let T be a coding of **D** in **E**, and let **R** and **S** be n-place relations on **D**. Show:

(a) $(\mathbf{R} \cap \mathbf{S})^T = \mathbf{R}^T \cap \mathbf{S}^T$
(b) $(\mathbf{R} \cup \mathbf{S})^T = \mathbf{R}^T \cup \mathbf{S}^T$
(c) $\mathbf{R} \subseteq \mathbf{S}$ if and only if $\mathbf{R}^T \subseteq \mathbf{S}^T$

Definition. Suppose that T is a coding of **D** in **E**. We call S the *decoding* of T if S assigns to each n-place relation **R** on **E** the n-place relation $\mathbf{R}^S$ on **D** consisting of those n-tuples $\langle d_1, \ldots, d_n \rangle$ with at least one T-code in **R**.

Figure 4.1 on the following page illustrates these notions. The coding T is indicated by the arrows from left to right. Then $\{a\}^T = \{p, q\}$ and $\{b, c\}^T = \{r, s, t, u\}$. The decoding of a set is the collection of all members of **D** that have arrows into the set. So $\{p, q, r, x\}^S = \{a, b\}$. Similarly for relations. For example,

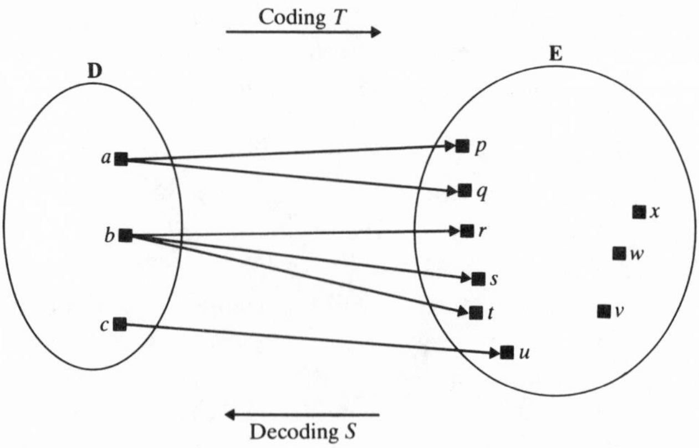

Figure 4.1

if $\mathbf{R}$ is the relation $\{\langle c, c\rangle, \langle a, c\rangle\}$ then $\mathbf{R}^T = \{\langle u, u\rangle, \langle p, u\rangle, \langle q, u\rangle\}$. If $\mathbf{S}$ is the relation $\{\langle p, q\rangle, \langle p, x\rangle, \langle p, r\rangle\}$ then $\mathbf{S}^S = \{\langle a, a\rangle, \langle a, b\rangle\}$.

As another example, suppose that T is the coding assigning to each set the collection of names for that set. Then the decoding S of T takes a collection of words, throws away all those that are not set names, and replaces those that are set names with the sets they name. The definition of decoding can be restated as follows. If T is a coding of $\mathbf{D}$ in $\mathbf{E}$, the decoding S assigns to each n-place relation $\mathbf{R}$ on $\mathbf{E}$ the n-place relation $\mathbf{R}^S$ on $\mathbf{D}$ such that $\langle d_1, \ldots, d_n\rangle \in \mathbf{R}^S$ if there are $e_1, \ldots, e_n \in \mathbf{E}$ with $\langle e_1, \ldots, e_n\rangle \in \mathbf{R}$ and $e_1 \in (d_1)^T, \ldots, e_n \in (d_n)^T$.

Exercise 4.2.2. Let T be a coding of $\mathbf{D}$ in $\mathbf{E}$, and let S be the decoding of T. Suppose that $\mathbf{R}$ is an n-place relation on $\mathbf{D}$. Show that $(\mathbf{R}^T)^S = \mathbf{R}$.

In the exercise above, $(\mathbf{R}^T)^S$ means we first code, then decode. In pictures, the exercise asserts the following diagram *commutes* in the sense that either way of "following the arrows" accomplishes the same thing.

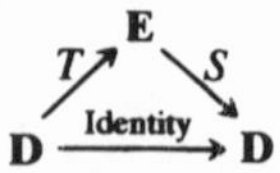

Definition. Let $\mathbf{D}$ and $\mathbf{E}$ be two domains with a coding T of $\mathbf{D}$ in $\mathbf{E}$, and with S as its decoding. Suppose Φ is an operator that maps k-place relations on $\mathbf{D}$ to n-place relations on $\mathbf{D}$. We define an operator Φ^T on $\mathbf{E}$ as follows. For each k-place relation $\mathbf{R}$ on $\mathbf{E}$, $\Phi^T(\mathbf{R}) = [\Phi(\mathbf{R}^S)]^T$.

Informally, Φ^T is to behave, in $\mathbf{E}$ on codes, the way Φ behaved in $\mathbf{D}$ on the original objects. To apply Φ^T to $\mathbf{R}$, first decode it, getting a relation on $\mathbf{D}$; apply

Φ to this relation, then code the result. Another way of putting it is, Φ^T is the operator that makes the following diagram commute.

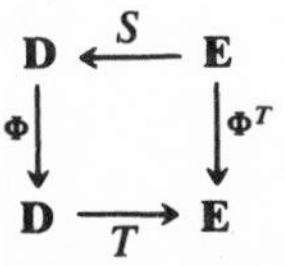

For example, again suppose T is the coding that assigns to each set the collection of names for that set. Let Φ be the "union" operator on sets in the following sense. If $\langle s_1, s_2\rangle$ is in the input relation then $s_1 \cup s_2$ is an output. More formally, $\Phi(\mathbf{R}) = \{s_1 \cup s_2 \mid \langle s_1, s_2\rangle \in \mathbf{R}\}$. Then Φ^T has the following behavior on words. Say that $\langle w_1, w_2\rangle$ is in the input supplied to Φ^T. If one of w_1 or w_2 is not a set name, this input plays no role. If both w_1 and w_2 are set names then the output includes all names for the set we get when we form the union of the set w_1 names and the set w_2 names.

Definition. Suppose Φ_1 and Φ_2 are two operators on $\mathbf{D}$ such that the output of Φ_2 and the input of Φ_1 "agree." That is, Φ_2 outputs n-place relations, and Φ_1 takes, as inputs, n-place relations. Then the *composition* $\Phi_1\Phi_2$ is defined, and its behavior is given by $(\Phi_1\Phi_2)(\mathbf{P}) = \Phi_1(\Phi_2(\mathbf{P}))$.

Informally, to apply $\Phi_1\Phi_2$ to $\mathbf{P}$, use $\mathbf{P}$ as input to Φ_2, take the output of this and use it as input for Φ_1. In diagrams,

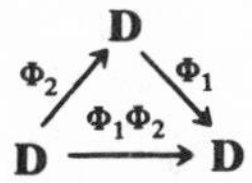

Exercise 4.2.3. Suppose that T is a coding of $\mathbf{D}$ in $\mathbf{E}$, with S its decoding. And let Φ_1 and Φ_2 be operators on $\mathbf{D}$, Φ_2 taking k-place relations to n-place relations, and Φ_1 taking n-place relations to m-place relations. (Then $\Phi_1\Phi_2$ is defined.) Show that $(\Phi_1\Phi_2)^T = \Phi_1^T\Phi_2^T$. *Hint*: Use Exercise 4.2.2, and consider the following diagram.

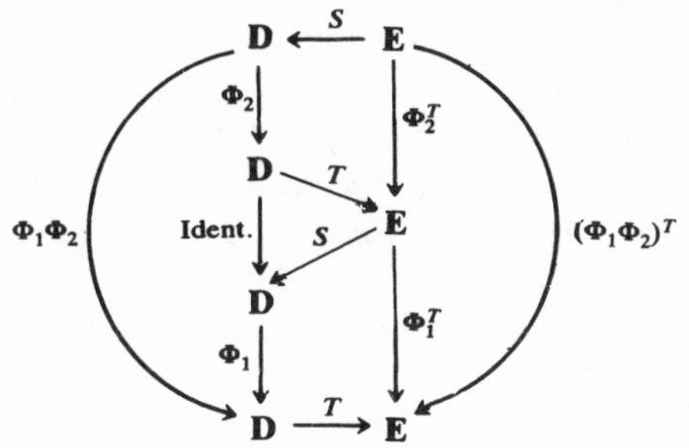

4.3 IMPLEMENTATIONS

Definition. Suppose that we have two data structures $\langle \mathbf{D}; \mathbf{R}_1, \ldots, \mathbf{R}_u \rangle$ and $\langle \mathbf{E}; \mathbf{S}_1, \ldots, \mathbf{S}_v \rangle$, and suppose that T is a coding of **D** in **E**. Also, let **EQUAL** be the equality relation on **D**, $\{\langle x, x \rangle \mid x \in \mathbf{D}\}$. We call T an *implementation* of $\langle \mathbf{D}; \mathbf{R}_1, \ldots, \mathbf{R}_u \rangle$ in $\langle \mathbf{E}; \mathbf{S}_1, \ldots, \mathbf{S}_v \rangle$ if each of the relations $\mathbf{R}_1^T, \ldots, \mathbf{R}_u^T$ and $\mathbf{EQUAL}^T$ is generated in EFS($\langle \mathbf{E}; \mathbf{S}_1, \ldots, \mathbf{S}_v \rangle$).

Exercise 4.3.1. Let T be a coding of **D** in **E**. Let **EQUAL** be the equality relation on **D**, and let **EQUIV** be the relation on **E** such that $\langle x, y \rangle \in \mathbf{EQUIV}$ just when x and y are T-codes for the same member of **D**. Show that $\mathbf{EQUAL}^T = \mathbf{EQUIV}$.

Thus, we have an implementation of one data structure in a second provided we can generate analogs of all the given relations of the first data structure, but on codes instead of on the original objects, and also we can discover when we have equivalent codes for the same object. For example, consider the coding T of nonnegative integers as binary tree number representatives, from Chapter 2, §2.3. Since each integer has just one code, equivalence of codes is trivial. Then Exercise 2.3.3 shows that T is actually an implementation of the data structure **integer** in **tree**(**A**). Similarly, in Chapter 2, §2.4, we gave an implementation of **integer** in **set**(**A**) (see Exercise 2.4.8). More examples will be given in later sections. There are examples where *no* implementation can exist. For instance, suppose we have a data structure $\langle \mathbf{D}; \mathbf{R}_1, \ldots, \mathbf{R}_k \rangle$ whose domain **D** is the set of *real* numbers. There is no implementation of this in the structure **integer**, because no coding can exist. This is a consequence of Cantor's famous result in set theory that says no one-one function from the reals to the integers can exist. An example where a coding exists but there is still no implementation will be given in §4.4.

In Chapter 2, §2.3, after creating an implementation of **integer** in **tree**(**A**), though it was not described as such at the time, you were asked to show the coded versions of LESS and LESS_OR_EQUAL were generated in EFS(**tree**(**A**)) (Exercise 2.3.4). The following fundamental result says such things can always be done.

Theorem 4.3.1. *Let* T *be an implementation of the data structure* $\langle \boldsymbol{D}; \boldsymbol{R}_1, \ldots, \boldsymbol{R}_{\mathrm{u}} \rangle$ *in* $\langle \boldsymbol{E}; \boldsymbol{S}_1, \ldots, \boldsymbol{S}_{\mathrm{v}} \rangle$.

1. *If* $\boldsymbol{R}$ *is a generated relation of* $EFS(\langle \boldsymbol{D}; \boldsymbol{R}_1, \ldots, \boldsymbol{R}_{\mathrm{u}} \rangle)$ *then* $\boldsymbol{R}^{\mathrm{T}}$ *is a generated relation of* $EFS(\langle \boldsymbol{E}; \boldsymbol{S}_1, \ldots, \boldsymbol{S}_{\mathrm{v}} \rangle)$.
2. *If* Φ *is an operator defined by a procedure in* $EFS(\langle \boldsymbol{D}; \boldsymbol{R}_1, \ldots, \boldsymbol{R}_{\mathrm{u}} \rangle)$ *then* Φ^{T} *is defined by a procedure in* $EFS(\langle \boldsymbol{E}; \boldsymbol{S}_1, \ldots, \boldsymbol{S}_{\mathrm{v}} \rangle)$.

The proof is long, and occupies most of the rest of this section, including several exercises. It is constructive; we say how to take a procedure of EFS($\langle \mathbf{D}; \mathbf{R}_1, \ldots, \mathbf{R}_u \rangle$) and turn it into one of EFS($\langle \mathbf{E}; \mathbf{S}_1, \ldots, \mathbf{S}_v \rangle$). We must then verify the resulting procedure does what it is supposed to do. We use models for this purpose, though a proof using traces could also be given. The argument for part 2 is similar to that for part 1 and is left to you.

The equality relation is always generated, no matter what the work space (Exercise 2.6.2). When working over **D** it will be convenient for us to assume it is explicitly among the given relations. This is for the sake of simplicity; it plays no essential role. When working with EFS($\langle \mathbf{D}; \mathbf{R}_1, \ldots, \mathbf{R}_u \rangle$) we use the work space $\mathbf{W_D}$ whose reserved identifiers are $R_1, \ldots, R_u$ and EQUAL, representing $\mathbf{R}_1, \ldots, \mathbf{R}_u$ and **EQUAL** respectively.

T is an implementation of $\langle \mathbf{D}; \mathbf{R}_1, \ldots, \mathbf{R}_u \rangle$ in $\langle \mathbf{E}; \mathbf{S}_1, \ldots, \mathbf{S}_v \rangle$. In particular, then, T is a coding of **D** in **E**; we let S be its decoding.

When working with EFS($\langle \mathbf{E}; \mathbf{S}_1, \ldots, \mathbf{S}_v \rangle$) it will be most convenient to use an *extension* of a basic work space. We describe the one we want. First, we use $S_1, \ldots, S_v$ as reserved identifiers representing the given relations $\mathbf{S}_1, \ldots, \mathbf{S}_v$, and we assume these are different from any of the reserved identifiers of $\mathbf{W_D}$. Next, T is an implementation so by definition each of $\mathbf{R}_1^T, \ldots, \mathbf{R}_u^T$ and $\mathbf{EQUAL}^T$ is generated. We assume these are also given relations of the work space we will use, and we take $R_1, \ldots, R_u$ and EQUAL to represent them. There should be no confusion in the dual role these identifiers play, since two separate data structures are involved. Finally, let CODE be an identifier not used so far, and let CODE(x) be given by EQUAL(x, x). CODE represents a generated relation of EFS($\langle \mathbf{E}; \mathbf{S}_1, \ldots, \mathbf{S}_v \rangle$); in fact, it clearly represents the set **CODE** of T codes. Now, when working over **E**, we use the following work space. The reserved identifiers are $S_1, \ldots, S_v$, $R_1, \ldots, R_u$, EQUAL, and CODE, and these represent the relations $\mathbf{S}_1, \ldots, \mathbf{S}_v$, $\mathbf{R}_1^T, \ldots, \mathbf{R}_u^T$, $\mathbf{EQUAL}^T$ and **CODE** respectively. We designate this work space by $\mathbf{W_E}$.

Suppose **REL** is a generated relation in EFS($\langle \mathbf{D}; \mathbf{R}_1, \ldots, \mathbf{R}_u \rangle$). Say that it is the output of the following procedure, written in work space $\mathbf{W_D}$.

REL (n):
(procedure body of REL).

We will write a procedure for $\mathbf{REL}^T$ in EFS($\langle \mathbf{E}; \mathbf{S}_1, \ldots, \mathbf{S}_v \rangle$), in the work space $\mathbf{W_E}$.

Without loss of generality we can assume none of the identifiers $S_1, \ldots, S_v$, or CODE have been used in any procedure statement of REL. Then because of the way we have reused the reserved identifiers of $\mathbf{W_D}$ in $\mathbf{W_E}$, a procedure statement X of REL is also acceptable in $\mathbf{W_E}$ (except for the possibility that X involves constants from **D**, a point we take up shortly). We might hope that X behaves in $\mathbf{W_E}$, on codes, the way X behaved in $\mathbf{W_D}$, on the objects being coded. But we must specify that we are not interested in the behavior of X in $\mathbf{W_E}$ on things that are not codes.

Definition. By the *code restriction* of X we mean

$$\text{CODE}(x_1) \rightarrow \text{CODE}(x_2) \rightarrow \ldots \rightarrow \text{CODE}(x_k) \rightarrow X$$

where $x_1, \ldots, x_k$ are all the variables of X (in some arbitrary order). We denote the code restriction of X by $X \restriction$CODE.

Objects in **D** may have several codes in **E**, but any one will do as well as any other. We can "tell" when we have equivalent codes, in $\mathbf{W_E}$, because $\mathbf{EQUAL}^T$ is given in the work space, represented by EQUAL.

Definition. Let IDENTIFIER be an identifier used in the procedure REL, representing a k-place relation. By an *equality statement* for IDENTIFIER we mean

$$\text{EQUAL}(x_1, y_1) \rightarrow \text{EQUAL}(x_2, y_2) \rightarrow \ldots \rightarrow \text{EQUAL}(x_k, y_k)$$
$$\rightarrow \text{IDENTIFIER}(x_1, \ldots, x_k) \rightarrow \text{IDENTIFIER}(y_1, \ldots, y_k).$$

Let X be some procedure statement of REL. X may involve constants, members of **D**. For each constant d occurring in X, choose some T code e for it, and let X' be the result of replacing, in X each constant by the code you chose for it. (If a constant has several codes, it will not matter which one you pick. Choose one arbitrarily.)

Now we are ready to write a procedure in work space $\mathbf{W_E}$ for $\mathbf{REL}^T$. We claim the following procedure will do.

REL (n):
    (translated procedure body of REL).

where (translated procedure body of REL) consists of the following items:

1. For each procedure statement X of REL, the procedure statement $X' \restriction$CODE, where X' is the result of replacing constants of X by arbitrary codes for them, as described above.
2. Equality statements for every unreserved identifier of REL.

Before we continue with the proof we present an example of things thus far. In Chapter 2, §2.2 the following procedure was written in the basic work space of EFS(**integer**).

SUM (3):
    SUM(x, 0, x);
    SUM(x, y, z) → SUC(y, u) → SUC(z, v) → SUM(x, u, v).

In Chapter 2, §2.4 number representatives were defined in **set**(**A**). Let T be the coding that assigns to each integer the set that is its number representative. Exercises 2.4.7 and 2.4.8 say that T is an implementation. We follow the method given above, and "translate" the procedure SUM from EFS(**integer**) to EFS(**set**(**A**)). We work in the expansion of the basic work space of EFS(**set**(**A**)), with the extra relations $\mathbf{SUC}^T$, $\mathbf{EQUAL}^T$, and **CODE**, represented by SUC, EQUAL, and CODE respectively. (*Note*: In Exercise 2.4.7 you were really asked to show **CODE** was generated. Since number representatives are unique,

$\mathbf{EQUAL}^T$ is just the equality relation on sets, restricted to those sets that are number representatives. That is, $\mathbf{EQUAL}^T$ is $\{\langle x, x\rangle \mid x \in \mathbf{CODE}\}$. It follows from this that $\mathbf{EQUAL}^T$ is generated too.)

Let X be the first procedure statement of SUM, $X = \text{SUM}(x, 0, x)$. This contains a constant, 0. The only number representative for this is the empty set, $\varnothing$, so take $X' = \text{SUM}(x, \varnothing, x)$. Then the code restriction, $X' \restriction \text{CODE}$, is $\text{CODE}(x) \to \text{SUM}(x, \varnothing, x)$. For the other procedure statement, since no constants are involved, the code restriction is just $\text{CODE}(x) \to \text{CODE}(y) \to \text{CODE}(z) \to \text{CODE}(u) \to \text{CODE}(v) \to \text{SUM}(x, y, z) \to \text{SUC}(y, u) \to \text{SUC}(z, v) \to \text{SUM}(x, u, v)$. Equality statements are not strictly necessary, since number representatives are unique, but we give them anyway, for illustration. There is only one; $\text{EQUAL}(x, u) \to \text{EQUAL}(y, v) \to \text{EQUAL}(z, w) \to \text{SUM}(x, y, z) \to \text{SUM}(u, v, w)$. So, the procedure we wind up with, on sets, is the following.

$$
\begin{array}{l}
\text{SUM}\,(3): \\
\quad \text{CODE}(x) \to \text{SUM}(x, \varnothing, x); \\
\quad \text{CODE}(x) \to \text{CODE}(y) \to \text{CODE}(z) \to \text{CODE}(u) \\
\qquad \to \text{CODE}(v) \to \text{SUM}(x, y, z) \to \text{SUC}(y, u) \\
\qquad \to \text{SUC}(z, v) \to \text{SUM}(x, u, v); \\
\quad \text{EQUAL}(x, u) \to \text{EQUAL}(y, v) \to \text{EQUAL}(z, w) \\
\qquad \to \text{SUM}(x, y, z) \to \text{SUM}(u, v, w).
\end{array}
$$

Now we return to the proof. We have given a translation method for procedures, we must show it works.

Definition. Let $\mathbf{I}$ be an interpretation in work space $\mathbf{W_D}$. By $\mathbf{I}T$ we mean the interpretation in $\mathbf{E}$ given by the following.

1. For each identifier other than CODE, $S_1, \ldots, S_v$ (which were not reserved in $\mathbf{W_D}$, nor which appeared in (procedure body of REL)), we set $\text{IDENTIFIER}^{\mathbf{I}T} = (\text{IDENTIFIER}^{\mathbf{I}})^T$. That is, interpret IDENTIFIER in $\mathbf{D}$ as $\mathbf{I}$ says, then transfer this relation to $\mathbf{E}$ using the coding T.
2. $\text{CODE}^{\mathbf{I}T}$ is simply the set of T codes in $\mathbf{E}$.
3. $(S_i)^{\mathbf{I}T} = \mathbf{S}_i$.

We can think of T as also being a map from interpretations to interpretations. Since an interpretation $\mathbf{I}$ in $\mathbf{W_D}$ must make $(R_i)^{\mathbf{I}} = \mathbf{R}_i$, then we have $(R_i)^{\mathbf{I}T} = (R_i^{\mathbf{I}})^T = (\mathbf{R}_i)^T$. Similarly for EQUAL. It follows that $\mathbf{I}T$ is an interpretation *in* $\mathbf{W_E}$.

Exercise 4.3.2. Suppose that $\mathbf{I}$ is an interpretation in $\mathbf{W_D}$ such that every procedure statement of (procedure body of REL) is true under $\mathbf{I}$. Show that every procedure statement of (translated procedure body of REL) is true in $\mathbf{W_E}$ under $\mathbf{I}T$. (Code restrictions play a role here.)

Definition. Let $\mathbf{J}$ be an interpretation in work space $\mathbf{W_E}$. By $\mathbf{J}S$ we mean the interpretation in $\mathbf{D}$ given by $\text{IDENTIFIER}^{\mathbf{J}S} = (\text{IDENTIFIER}^{\mathbf{J}})^S$.

Then S too can be thought of as a map from interpretations to interpretations. For an interpretation $\mathbf{J}$ *in* $\mathbf{W_E}$, for each reserved identifier R_i of $\mathbf{W_D}$ we have, by definition of $\mathbf{W_E}$, $(R_i)^{\mathbf{J}} = \mathbf{R}_i^T$. Then $(R_i)^{\mathbf{J}S} = (R_i^{\mathbf{J}})^S = (\mathbf{R}_i^T)^S = \mathbf{R}_i$, by Exercise 4.2.2. Similarly for EQUAL. This means that $\mathbf{J}S$ is an interpretation *in* $\mathbf{W_D}$.

Exercise 4.3.3. Suppose $\mathbf{J}$ is an interpretation in $\mathbf{W_E}$ such that every procedure statement of (translated procedure body of REL) is true under $\mathbf{J}$. Show that every procedure statement of (procedure body of REL) is true in $\mathbf{W_D}$ under $\mathbf{J}S$.

Because of Exercise 4.2.2, if $\mathbf{I}$ is an interpretation in $\mathbf{W_D}$, then $(\mathbf{I}^T)^S = \mathbf{I}$. We need a similar result with S and T reversed. We have the following weak version.

Exercise 4.3.4.

(a) Suppose that $\mathbf{P}$ is a relation on $\mathbf{E}$ that is closed under code equivalence. That is, if $\langle a_1, \ldots, a_n \rangle$ and $\langle b_1, \ldots, b_n \rangle$ are T-codes for the same n-tuple over $\mathbf{D}$, and if $\langle a_1, \ldots, a_n \rangle \in \mathbf{P}$, then $\langle b_1, \ldots, b_n \rangle \in \mathbf{P}$. Show that $(\mathbf{P}^S)^T \subseteq \mathbf{P}$.

(b) Suppose $\mathbf{J}$ is an interpretation in $\mathbf{W_E}$ such that the equality statement for IDENTIFIER is true under $\mathbf{J}$. Show that $((\text{IDENTIFIER}^{\mathbf{J}})^S)^T \subseteq \text{IDENTIFIER}^{\mathbf{J}}$.

Now we have everything we need to finish the proof. Let $\mathbf{M}$ be the minimal model for the procedure REL in $\mathbf{W_D}$. By Proposition 1.7.2, the procedure output, **REL**, is $\text{REL}^{\mathbf{M}}$. Likewise, let $\mathbf{N}$ be the minimal model for the translated version of REL in $\mathbf{W_E}$. By Proposition 1.7.2 again, the output of this procedure is $\text{REL}^{\mathbf{N}}$. So what we must show is that $(\text{REL}^{\mathbf{M}})^T = \text{REL}^{\mathbf{N}}$.

By Exercise 4.3.2, since $\mathbf{M}$ is a model for REL, then $\mathbf{M}T$ is a model for the translated REL. Since $\mathbf{N}$ is the *minimal* such model, $\mathbf{N} \leq \mathbf{M}T$ (Exercise 3.9.4). But then $\text{REL}^{\mathbf{N}} \subseteq \text{REL}^{\mathbf{M}T} = (\text{REL}^{\mathbf{M}})^T$. Further, since $\mathbf{N}$ is a model for the translated REL, by Exercise 4.3.3, $\mathbf{N}S$ is a model for REL itself. Since $\mathbf{M}$ is the minimal such model, $\mathbf{M} \leq \mathbf{N}S$. Then $\text{REL}^{\mathbf{M}} \subseteq \text{REL}^{\mathbf{N}S} = (\text{REL}^{\mathbf{N}})^S$. It follows by Exercise 4.3.4 that $(\text{REL}^{\mathbf{M}})^T \subseteq ((\text{REL}^{\mathbf{N}})^S)^T \subseteq \text{REL}^{\mathbf{N}}$.

This concludes the proof of Theorem 4.3.1.

Exercise 4.3.5. Prove part 2 of Theorem 4.3.1.

Exercise 4.3.6. Show that, in the proof above, in translating the procedure REL from $\text{EFS}(\langle \mathbf{D}; \mathbf{R}_1, \ldots, \mathbf{R}_n \rangle)$ to $\text{EFS}(\langle \mathbf{E}; \mathbf{S}_1, \ldots, \mathbf{S}_v \rangle)$ it wasn't necessary to add equality statements for all the unreserved identifiers, but only for the output identifier.

Definition. Let T be a coding of $\mathbf{D}$ in $\mathbf{E}$, and U be a coding of $\mathbf{E}$ in $\mathbf{F}$. We define a *composition* TU as follows. For each $d \in \mathbf{D}$, $d^{(TU)} = (d^T)^U$. That is, $d^{(TU)}$ is the collection of all U-codes for T-codes for d; $d^{(TU)} = \{x \in \mathbf{F} \mid x \in y^U \text{ for some } y \in d^T\}$.

Exercise 4.3.7. Verify that the composition of codings is a coding.

Exercise 4.3.8. Suppose that T is a coding of $\mathbf{D}$ in $\mathbf{E}$ and U is a coding of $\mathbf{E}$ in $\mathbf{F}$. Let $\mathbf{R}$ be an n-place relation on $\mathbf{D}$. Show that $\mathbf{R}^{(TU)} = (\mathbf{R}^T)^U$.

Corollary 4.3.2. *Suppose that T is an implementation of* $\langle \boldsymbol{D}; \boldsymbol{R}_1, \ldots, \boldsymbol{R}_u \rangle$ *in* $\langle \boldsymbol{E}; \boldsymbol{S}_1, \ldots, \boldsymbol{S}_v \rangle$ *and* U *is an implementation of* $\langle \boldsymbol{E}; \boldsymbol{S}_1, \ldots, \boldsymbol{S}_v \rangle$ *in* $\langle \boldsymbol{F}; \boldsymbol{T}_1, \ldots, \boldsymbol{T}_w \rangle$. *Then* TU *is an implementation of* $\langle \boldsymbol{D}; \boldsymbol{R}_1, \ldots, \boldsymbol{R}_u \rangle$ *in* $\langle \boldsymbol{F}; \boldsymbol{T}_1, \ldots, \boldsymbol{T}_w \rangle$.

Proof. Exercise 4.3.7 says that TU is a coding. If $\mathbf{R}_i$ is a given relation of $\langle \mathbf{D}; \mathbf{R}_1, \ldots, \mathbf{R}_u \rangle$, we want $\mathbf{R}_i^{(TU)} = (\mathbf{R}_i^T)^U$ to be generated in $\langle \mathbf{F}; \mathbf{T}_1, \ldots, \mathbf{T}_w \rangle$. Since T is an implementation, $\mathbf{R}_i^T$ is generated in $\langle \mathbf{E}; \mathbf{S}_1, \ldots, \mathbf{S}_v \rangle$. Then, by Theorem 4.3.1, $(\mathbf{R}_i^T)^U$ is generated in $\langle \mathbf{F}; \mathbf{T}_1, \ldots, \mathbf{T}_w \rangle$. Similarly for **EQUAL**. Then, by definition, TU is an implementation.

4.4 CONSERVATIVE IMPLEMENTATIONS

Definition. Let T be an implementation of $\langle \mathbf{D}; \mathbf{R}_1, \ldots, \mathbf{R}_u \rangle$ in $\langle \mathbf{E}; \mathbf{S}_1, \ldots, \mathbf{S}_v \rangle$. We say that T is *conservative* if, for each relation $\mathbf{R}$ on $\mathbf{D}$, $\mathbf{R}$ is generated in $\langle \mathbf{D}; \mathbf{R}_1, \ldots, \mathbf{R}_u \rangle$ if and only if $\mathbf{R}^T$ is generated in $\langle \mathbf{E}; \mathbf{S}_1, \ldots, \mathbf{S}_v \rangle$.

Proposition 4.4.1. *Suppose that* $\langle \boldsymbol{D}; \boldsymbol{R}_1, \ldots, \boldsymbol{R}_u \rangle$ *has a conservative implementation in* $\langle \boldsymbol{E}; \boldsymbol{S}_1, \ldots, \boldsymbol{S}_v \rangle$ *which, in turn, has a conservative implementation in* $\langle \boldsymbol{F}; \boldsymbol{T}_1, \ldots, \boldsymbol{T}_w \rangle$. *Then* $\langle \boldsymbol{D}; \boldsymbol{R}_1, \ldots, \boldsymbol{R}_u \rangle$ *has a conservative implementation in* $\langle \boldsymbol{F}; \boldsymbol{T}_1, \ldots, \boldsymbol{T}_w \rangle$.

Exercise 4.4.1. Prove Proposition 4.4.1.

Not all implementations are conservative. As a trivial example, let **weakinteger** be the data structure whose domain is the set of nonnegative integers, but with *no* relations given. There is an obvious implementation of **weakinteger** in **integer**: Code each integer by itself. But we can do much more in EFS(**integer**) than we can in EFS(**weakinteger**). As a less trivial example, consider an implementation of a data structure of *unordered* binary trees in one of *ordered* binary trees where we code an unordered tree by a tree that is "like" it but with an order to the branches chosen arbitrarily. If this were a conservative implementation we would have a way of ordering branches in the data structure of unordered trees, which is not generally possible.

Our main concern in this section is how to verify when we *do* have a conservative implementation. We supply a simple test; the remainder of the chapter is devoted to examples of the test applied. We do not claim that the test covers all cases in which a conservative implementation exists; but it is sufficient for our purposes. Suppose we have two data structures $\langle \mathbf{D}; \mathbf{R}_1, \ldots, \mathbf{R}_u \rangle$ and

$\langle \mathbf{E}; \mathbf{S}_1, \ldots, \mathbf{S}_v \rangle$, and suppose each has an implementation in the other. Say that T is an implementation of the **D**-structure in the **E**-structure, while U is an implementation in the other direction. Then the composition TU is also an implementation, of $\langle \mathbf{D}; \mathbf{R}_1, \ldots, \mathbf{R}_u \rangle$ in *itself.*

Definition. Let T be an implementation of $\langle \mathbf{D}; \mathbf{R}_1, \ldots, \mathbf{R}_u \rangle$ in $\langle \mathbf{E}; \mathbf{S}_1, \ldots, \mathbf{S}_v \rangle$, and let U be an implementation of $\langle \mathbf{E}; \mathbf{S}_1, \ldots, \mathbf{S}_v \rangle$ in $\langle \mathbf{D}; \mathbf{R}_1, \ldots, \mathbf{R}_u \rangle$. For each $d \in \mathbf{D}$, there is a set $d^{(TU)} \subseteq \mathbf{D}$ of TU-codes for d. We say the *TU coding is generated in* $\langle \mathbf{D}; \mathbf{R}_1, \ldots, \mathbf{R}_u \rangle$ if the 2-place relation $e \in d^{(TU)}$ is generated in EFS($\langle \mathbf{D}; \mathbf{R}_1, \ldots, \mathbf{R}_u \rangle$).

Theorem 4.4.2. *Suppose that* T *is an implementation of* $\langle \boldsymbol{D}; \boldsymbol{R}_1, \ldots, \boldsymbol{R}_u \rangle$ *in* $\langle \boldsymbol{E}; \boldsymbol{S}_1, \ldots, \boldsymbol{S}_v \rangle$ *and* U *is an implementation of* $\langle \boldsymbol{E}; \boldsymbol{S}_1, \ldots, \boldsymbol{S}_v \rangle$ *in* $\langle \boldsymbol{D}; \boldsymbol{R}_1, \ldots, \boldsymbol{R}_u \rangle$, *such that the* TU *coding is generated in* $\langle \boldsymbol{D}; \boldsymbol{R}_1, \ldots, \boldsymbol{R}_u \rangle$. *Then* T *is a conservative implementation.*

Proof. Let $\mathbf{R}$ be an n-place relation on $\mathbf{D}$. By Theorem 4.3.1, if $\mathbf{R}$ is generated in $\langle \mathbf{D}; \mathbf{R}_1, \ldots, \mathbf{R}_u \rangle$ then $\mathbf{R}^T$ is generated in $\langle \mathbf{E}; \mathbf{S}_1, \ldots, \mathbf{S}_v \rangle$. Conversely, suppose $\mathbf{R}^T$ is generated in $\langle \mathbf{E}; \mathbf{S}_1, \ldots \mathbf{S}_v \rangle$. By Theorem 4.3.1 again, $(\mathbf{R}^T)^U$ is generated in $\langle \mathbf{D}; \mathbf{R}_1, \ldots, \mathbf{R}_u \rangle$, that is, by Exercise 4.3.8, $\mathbf{R}^{(TU)}$ is generated. But the relation $\mathbf{R}$ can be recovered from $\mathbf{R}^{(TU)}$ because

$$\mathbf{R}(x_1, \ldots, x_n) \Leftrightarrow (\exists y_1) \ldots (\exists y_n)[\mathbf{R}^{(TU)}(y_1, \ldots, y_n) \wedge y_1 \in x_1^{(TU)} \wedge \ldots \wedge y_n \in x_n^{(TU)}].$$

By our assumptions, and the closure of generated relations under $\wedge$ and $\exists$, $\mathbf{R}$ must be generated.

Exercise 4.4.2. Consider the data structure **integer** = $\langle \mathbf{N}; \mathbf{SUC} \rangle$. Let $\mathbf{R}$ be some relation on $\mathbf{N}$, and let T be an implementation of $\langle \mathbf{N}; \mathbf{SUC}, \mathbf{R} \rangle$ in **integer**.

(a) Since T, as a coding, is from $\mathbf{N}$ to itself, the relation *y is a T-code of x* is a relation on $\mathbf{N}$. Show that it must be generated in **integer**.
(b) Use part (a) and show that $\mathbf{R}$ must be generated in **integer**.

It will be shown in Chapter 6, §6.8 that there are sets of integers that are not generated in EFS(**integer**). Let $\mathbf{R}$ be such a set. Then the data structure $\langle \mathbf{N}; \mathbf{SUC}, \mathbf{R} \rangle$ can have no implementation in **integer** because of the exercise above. There are, of course, lots of codings possible, but none can be an *implementation*.

Theorem 4.4.2 can be given in a stronger form, which we leave to you in exercises.

Exercise 4.4.3. Suppose that T is an implementation of $\langle \mathbf{D}; \mathbf{R}_1, \ldots, \mathbf{R}_u \rangle$ in $\langle \mathbf{E}; \mathbf{S}_1, \ldots, \mathbf{S}_v \rangle$, and S is the decoding of T. Let $\mathbf{S}$ be a generated relation of $\langle \mathbf{E}; \mathbf{S}_1, \ldots, \mathbf{S}_v \rangle$. Show that $\mathbf{S}^{ST}$ is also generated.

Definition. Let T be an implementation of $\langle \mathbf{D}; \mathbf{R}_1, \ldots, \mathbf{R}_u \rangle$ in $\langle \mathbf{E}; \mathbf{S}_1, \ldots, \mathbf{S}_v \rangle$, and let S be its decoding. We say that T is *strongly conservative* if, for each generated relation $\mathbf{S}$ on $\mathbf{E}$, $\mathbf{S}^S$ is generated in $\langle \mathbf{D}; \mathbf{R}_1, \ldots, \mathbf{R}_u \rangle$.

Exercise 4.4.4. Show that a strongly conservative implementation is conservative.

Proposition 4.4.3. *Under the same hypotheses as Theorem* 4.4.2, T *is a strongly conservative implementation.*

Exercise 4.4.5. Prove Proposition 4.4.3.

Exercise 4.4.6. Show that, under the hypotheses of Theorem 4.4.2, the operator Φ is defined by a procedure in EFS($\langle \mathbf{D}; \mathbf{R}_1, \ldots, \mathbf{R}_u \rangle$) if and only if Φ^T is defined by a procedure in EFS($\langle \mathbf{E}; \mathbf{S}_1, \ldots, \mathbf{S}_v \rangle$).

4.5 DIFFERENT SIZED ALPHABETS

This section is the first of several establishing the existence of conservative implementations. We show that if L and M are two finite alphabets with at least two letters each, then each of **str**(L) and **str**(M) has a conservative implementation in the other. The case of a one-letter alphabet is postponed until the next section.

In what follows, we assume that when we have two alphabets of different sizes, one is a subalphabet of the other. This assumption is not necessary, but it will simplify the presentation. For the rest of this section, L and M are fixed alphabets of at least two letters each, with L a subalphabet of M. Specifically, $L = \{a_1, \ldots, a_n\}$ and $M = \{a_1, \ldots, a_n, a_{n+1}, \ldots, a_m\}$.

Implementing str(L) in str(M). This direction is simple because every word over L is also a word over M. We use a coding T of L^* in M^* that codes each word over L by itself.

Exercise 4.5.1. Let $\mathbf{CODE}_L$ be the set of words over M that are made up of letters of L only. (That is, $\mathbf{CODE}_L = L^*$.) Show that $\mathbf{CODE}_L$ is generated in EFS(**str**(M)).

Exercise 4.5.2. Suppose we are working with EFS($\langle \mathbf{D}; \mathbf{R}_1, \ldots, \mathbf{R}_k \rangle$). Let $\mathbf{P} \subseteq \mathbf{D}$. For an n-place relation $\mathbf{R}$ on $\mathbf{D}$, define the *P-restriction* of $\mathbf{R}$ by

$\mathbf{R}\restriction\mathbf{P} = \{\langle x_1, \ldots, x_n\rangle \in \mathbf{R} \mid x_1 \in \mathbf{P}$ and $\ldots$ and $x_n \in \mathbf{P}\}$. Show that if both **R** and **P** are generated, so is $\mathbf{R}\restriction\mathbf{P}$.

Now it is easy to see that $(\mathbf{EQUAL}_L)^T = \mathbf{EQUAL}_M \restriction \mathbf{CODE}_L$ and $(\mathbf{CON}_L)^T = \mathbf{CON}_M \restriction \mathbf{CODE}_L$. ($\mathbf{EQUAL}_L$ is the equality relation on L^*, similarly for $\mathbf{EQUAL}_M$.) It follows from the two exercises above that these relations are generated in EFS(**str**(M)), so T is an implementation.

Implementing str(M) in str(L). $L = \{a_1, \ldots, a_n\}$ has at least two letters; let us write a_1 and a_2 as a and $'$ respectively. Then a'''' is a member of L^*, for instance. Define a coding U of M^* in L^* as follows. The U-code of the letter a_i is $a''\ldots'$, where we have written i primes. Next, we code words of M^* by coding each letter. If $c_1c_2\ldots c_t$ is a word of M^*, where each c_i is a letter, then $(c_1c_2\ldots c_t)^U = c_1^U c_2^U \ldots c_t^U$. For example, the U-code of the word $a_3a_5a_4$ is the word $a'''a'''''a''''$.

Exercise 4.5.3. Let $\mathbf{CODE}_M$ be the set of words over L that are U-codes for words over M. Show that $\mathbf{CODE}_M$ is generated in EFS(**str**(L)).

Again, it is easy to see that $(\mathbf{EQUAL}_M)^U = \mathbf{EQUAL}_L \restriction \mathbf{CODE}_M$ and $(\mathbf{CON}_M)^U = \mathbf{CON}_L \restriction \mathbf{CODE}_M$. It follows that these are generated in EFS(**str**(L)), so U is an implementation.

T Is Conservative. We show that the TU coding is generated in EFS(**str**(L)), then we apply Theorem 4.4.2. The following procedure in EFS(**str**(L)) will do. (Recall that a_1 and a_2 are the same things as a and $'$; we write them both ways for clarity only. Also recall that the alphabet L is $\{a_1, \ldots, a_n\}$.

```
TU_CODE (2):
    TU_CODE(Δ, Δ);
    TU_CODE(a₁, a');
    TU_CODE(a₂, a");
    :
    TU_CODE(aₙ, a"...');
    TU_CODE(x, y)→TU_CODE(u, v)→CON(x, u, w)
        →CON(y, v, z)→TU_CODE(w, z).
```

Exercise 4.5.4. Show that U is conservative.

Exercise 4.5.5. Let $K = \{a, b\}$. Define two "successor" relations on K^* as follows.

$\mathbf{SUCA}(x, y) \Leftrightarrow y = xa$
$\mathbf{SUCB}(x, y) \Leftrightarrow y = xb$

Thus **SUCA**(*abb, abba*) for example. Let *id* be the identity coding on K^* that codes each word by itself.

(a) Show that *id* is an implementation of **str**(K) in $\langle K^*; \mathbf{SUCA}, \mathbf{SUCB}\rangle$.
(b) Show that *id* is an implementation of $\langle K^*; \mathbf{SUCA}, \mathbf{SUCB}\rangle$ in **str**(K).
(c) Show that *id* is a conservative implementation in each of parts (a) and (b).

By the results of this section (and the next for one-letter alphabets), the exact size of an alphabet is not very important. If we want to show that some relation **R** on L^* is generated in EFS(**str**(L)) it may be convenient to move to a larger alphabet M and show that **R** is generated in EFS(**str**(M)). It follows that **R** is generated in EFS(**str**(L)) after all. We do not consider the question of whether adding extra letters affects *efficiency* of generation. We are only concerned with the *possibility* of generation.

4.6 NUMBERS AND WORDS

We show that there are conservative implementations between **integer** and **str**(L) for any finite alphabet L.

Exercise 4.6.1. Show that each of **integer** and **str**($\{1\}$) has a conservative implementation in the other.

In the previous section we showed that string data structures have conservative implementations in each other provided the alphabets have at least two letters. The exercise above says there is a similar connection between **integer** and **str**($\{1\}$). If we show each of **integer** and **str**($\{0, 1\}$) has a conservative implementation in the other, Proposition 4.4.1 will give us conservative implementations between any two string data structures, and between any string data structure and **integer**. This is what we devote the rest of the section to.

Implementing Integer in str$\{0, 1\}$. We want a coding of numbers by words over $\{0, 1\}$. There are many ways of doing this. For instance, we could code the integer n by its base 2 representation. We showed, as an example in Chapter 1, §1.5, that the successor relation on base 2 number names was generated in EFS(**str**($\{0, 1\}$)). But it will be more convenient to pick a simpler, though less efficient, representation.

Definition. For an integer n, n^T is 11...1, where we have written n 1's.

Exercise 4.6.2. Show that T is an implementation of **integer** in **str**($\{0, 1\}$).

Implementing str$\{0, 1\}$ in Integer. A word over $\{0, 1\}$ can be thought of as a base 2 name, so we might be tempted to code it by the integer it names. But this would code 101 and 0101 by the same integer; leading 0's would be ignored.

There are several ways around this. We adopt a simple device: Prefix each word with a 1 before thinking of it as a base 2 name. Specifically, we define U as follows.

Definition. For a word w over $\{0, 1\}$, $w^U = (1w)_2$ where $1w$ is the result of concatenating 1 and w, and $(1w)_2$ is the integer that $1w$ names in base 2.

We remind you that when items have unique codes we allow ourselves to write $c = w^U$ instead of $c \in w^U$. For examples, $(01)^U = (101)_2 = (5)_{10}$ and $(001)^U = (1001)_2 = (9)_{10}$.

Suppose that x and y are words over $\{0, 1\}$, and say that $x^U = a$ and $y^U = b$. If n is a number, write len(n) for the length of the base 2 name of n. We claim

$$(xy)^U = (b - 2^{\text{len}(b)-1}) + (a * 2^{\text{len}(b)-1}).$$

For example, if $x = 01$ and $y = 001$, then $(xy)^U = (01001)^U = (101001)_2 = (41)_{10}$. But also, writing everything in base 10 notation, $x^U = 5$, $y^U = 9$ and len(9) = 4. And $(9 - 2^{4-1}) + (5 * 2^{4-1}) = 1 + 40 = 41$. This looks mysterious, but only because we presented the work in base 10 notation. Consider it again, but this time in base 2 notation. Then $x^U = a = 101$ and $y^U = b = 1001$. Now $2^{\text{len}(b)-1}$ is simply 1000, and so $b - 2^{\text{len}(b)-1}$ is b with its leading 1 removed, while $a * 2^{\text{len}(b)-1}$ is a shifted left three places to "make room for" b. Then $(b - 2^{\text{len}(b)-1}) + (a * 2^{\text{len}(b)-1}) = (001) + (101000) = 101001$ which, of course, is $(xy)^U$.

Exercise 4.6.3. Show that U is an implementation by showing:

(a) Let **CON** be the concatenation relation on words over $\{0, 1\}$. Show that $\mathbf{CON}^U$ is generated in EFS(**integer**). *Hint*: See Exercise 2.2.1.
(b) Let **CODE** be the set of U-codes for words over $\{0, 1\}$. Show that **CODE** is generated in EFS(**integer**).
(c) Let **EQUAL** be the equality relation on $\{0, 1\}^*$. Show that $\mathbf{EQUAL}^U$ is generated in EFS(**integer**).

T Is Conservative. By Theorem 4.4.2 it is enough to show that the TU coding is generated in EFS(**integer**). But, this follows from Exercise 2.2.1 and the fact that for an integer n,

$$\begin{aligned} n^{TU} &= (n^T)^U \\ &= (11\ldots1)^U \qquad n \text{ 1's} \\ &= (111\ldots1)_2 \qquad n+1 \text{ 1's} \\ &= 2^{n+1} - 1. \end{aligned}$$

U Is Conservative. We must show that the UT coding is generated in

EFS(**str**($\{0, 1\}$)). Let w be a word over $\{0, 1\}$. Then

$$\begin{aligned} w^{UT} &= (w^U)^T \\ &= ((1w)_2)^T \\ &= 11\ldots1 \text{ with } (1w)_2 \text{ 1's.} \end{aligned}$$

Exercise 4.6.4. Write a procedure over EFS(**str**($\{0, 1\}$)) whose output is the following relation: x is any word, and y is a string of $(x)_2$ 1's.

Exercise 4.6.5. Finish the proof that U is conservative.

Exercise 4.6.6. (Cantor's pairing function). Let $\mathbf{N} = \{0, 1, 2, \ldots\}$, and define a function $J : \mathbf{N} \times \mathbf{N} \to \mathbf{N}$ by

$$J(x, y) = \frac{(x+y)(x+y+1)}{2} + y$$

Show the following.

(a) $J(n+1, 0) = J(0, n) + 1$
(b) if $n \neq 0$, $J(n-1, k+1) = J(n, k) + 1$
(c) J is onto
(d) (generalizing a) if $n > k$ then $J(n, 0) > J(0, k)$ because

$$J(n, 0) = J(0, k) + \frac{n(n+1) - (k+1)(k+2)}{2} + 1$$

(e) (generalizing b) if $a + b = c + d$ and $b > d$ then $J(a, b) > J(c, d)$ because $J(a, b) = J(c, d) + (b - d)$
(f) If $a + b > c + d$ then $J(a, b) > J(c, d)$ (Hint: Use part (d).)
(g) J is $1-1$
(h) The graph of J is generated in EFS(**integer**).

Exercise 4.6.7. (Continuing Exercise 4.6.6). Define two binary relations on $\mathbf{N} \times \mathbf{N}$:

XSUC($\langle a, b\rangle, \langle c, d\rangle$)$\Leftrightarrow a + 1 = c$ and $b = d$.
YSUC($\langle a, b\rangle, \langle c, d\rangle$)$\Leftrightarrow a = c$ and $b + 1 = d$.

Let **integerpairs** be the data structure $\langle \mathbf{N} \times \mathbf{N}$; **XSUC**, **YSUC**$\rangle$. Also, define two codings as follows.

T is a coding of $\mathbf{N} \times \mathbf{N}$ in $\mathbf{N}$ given by: the only T code of $\langle a, b\rangle$ is the number $J(a, b)$, where J is the function defined in Exercise 4.6.6. (Part (g) says this is a coding.)

U is a coding of $\mathbf{N}$ in $\mathbf{N} \times \mathbf{N}$ given by: the only U code of n is $\langle n, 0\rangle$. Show the following.

(a) U is an implementation of **integer** in **integerpairs**.
(b) T is an implementation of **integerpairs** in **integer**.
(c) T is a conservative implementation.
(d) U is a conservative implementation.

One can think of Exercise 4.6.7 as saying that arrays of length 2 can be simulated in **integer**. This can be extended to arrays of arbitrary length. Also, the result generalizes to many other data structures besides **integer** though we do not consider this here. (See Exercises 4.8.5 and 4.8.6, however.)

4.7 TREES AND WORDS

If there is to be a coding of trees by words it is necessary that the atoms be codable by words. As defined in Chapter 2, §2.3, any set of things can serve as atoms (except for a set of trees). If the atoms are the real numbers, there are results from set theory that say a coding of real numbers by finite words over a finite alphabet is not possible. There are "too many" real numbers to have names. In order to present basic ideas without getting involved in these kinds of difficulties we only consider the case of a *finite* set of atoms, which we can think of as being letters, although the results hold more generally.

For the rest of this section, L is a fixed finite alphabet with at least two members, which we also take as a set of atoms. When working with words it will be convenient to use a somewhat larger alphabet, although as we showed in §4.5 and §4.6, this does not really matter. Let **(**,**)** and **.** be three letters not already in L, and let L' be the alphabet L with these extra letters added. We will show there are conservative implementations between **str**(L') and **tree**(L).

Implementing Tree(L) in str(L'). The implementation we use in this direction is essentially the dot notation of LISP.

Definition. Define a coding T recursively as follows. For atoms a, let $a^T = a$. And, if T has already been defined on A and B, then let the T-code of the tree

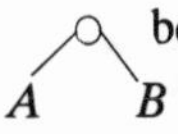

be the word $(A^T.B^T)$.

For example, assuming that a, b, c, d, and e are atoms, the T-code of the tree

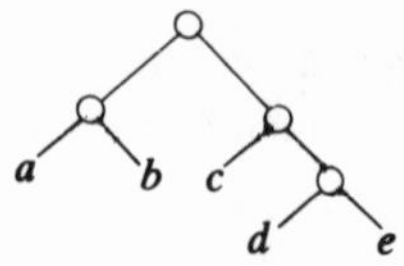

is the word **((**a **.** b**)** **.** **(**c **.** **(**d **.** e**)))**.

Exercise 4.7.1. Let **TCODE** be the set of T-codes for trees. Show that **TCODE** is generated in EFS(**str**(L')).

Exercise 4.7.2. Let **IMSUB** be the relation on **tree**(L) consisting of triples $\langle x, y, z \rangle$ where x is a tree and y and z are its immediate left and right subtrees. Show that $\mathbf{IMSUB}^T$ is generated in EFS(**str**(L')).

Exercise 4.7.3. Complete the proof that T is an implementation.

Implementing str(L') in tree(L). We need representatives for the letters of L' as trees. This is not the actual coding, but must be defined first in order that the code definition can be given. Recall, we said that L had at least two letters; let a and b be two distinct members of L.

Definition. Each letter of L is its own *tree representative.* And further

the tree representative of	is
(	
•	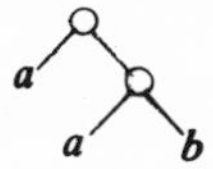
)	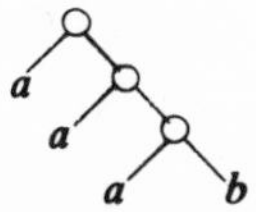

Definition. The coding U is given as follows. The U-code of Δ, the empty word, is the trivial tree consisting of just the atom a. If $c_1c_2 \ldots c_n$ is a word over L', where each c_i is a letter, and r_i is the tree representative of c_i, then $(c_1c_2 \ldots c_n)^U$ is the tree.

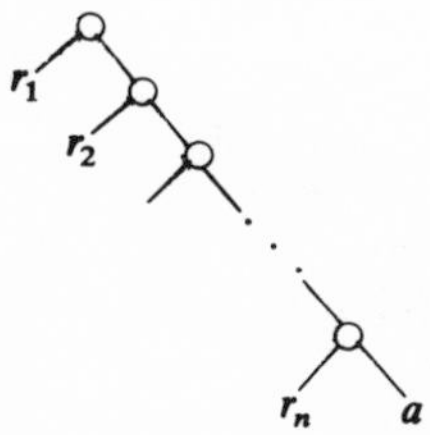

Note that if c is a letter of L, the tree representative of c is c itself, while c^U is 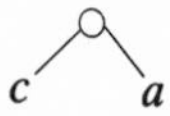

Exercise 4.7.4. Let **UCODE** be the set of U-codes for words over L'. Show that **UCODE** is generated in EFS(**tree**(L)).

We must show that **CON**^{U} is generated in EFS(**tree**(L)) where **CON** is the concatenation relation on words over L'. The procedure below will serve, but a few words of explanation may help in understanding it. For the first procedure statement, recall that the trivial tree a is the U-code for the empty word. For the second procedure statement, if x and y are words over L' and u is a letter, using juxtaposition to indicate concatenation, if $xy = z$, then $(ux)y = uz$. In the procedure, UCODE represents the set of U-codes (you wrote a procedure for it in Exercise 4.7.4) and REP represents the set of tree representatives for letters of L', a finite set, hence trivially generated.

U_CON (3):
  UCODE(x)→U_CON(a, x, x);
  U_CON(x, y, z)→REP(u)→IMSUB(r, u, x)
    →IMSUB(s, u, z)→U_CON(r, y, s).

Exercise 4.7.5. Complete the proof that U is an implementation.

U Is Conservative

Definition. If c is a letter of L', c has a tree representative, c' which, in turn has a T-code, c'', for it. Let us call c'' the *double representative* of c.

For example, if c is a letter of L, its double representative is c itself. If c is, say, **(**, the double representative is **(**a**.**b**)**. Since L' is finite, the relation of being the double representative is generated in EFS(**str**(L')). Say $qrst$ is a word over L', where q, r, s, and t are letters. Let q'', r'', s'', and t'' be their double representatives. It is easy to see that the only UT-code of $qrst$ is the word **(**q''**.(**r''**.(**s''**.(**t''**.**a**))))**.

Exercise 4.7.6. Show the UT coding is generated in EFS(**str**(L')).

T Is Conservative. Suppose we write **(**, **.**, and **)** for the tree representatives of (, ., and). And this time, let q, r, s, and t be letters of L. It is easy to see that the only TU-code of the tree

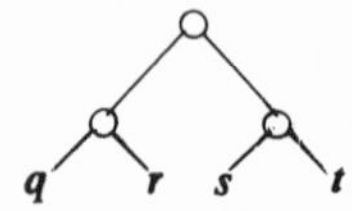

is the tree

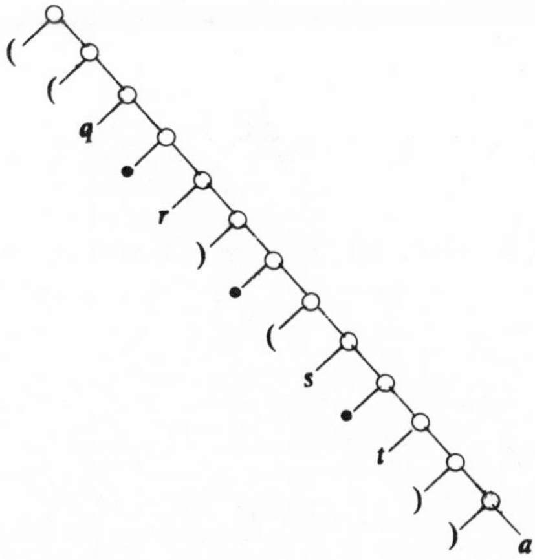

Exercise 4.7.7. Show that the *TU* coding is generated in EFS(**tree**(L)).

4.8 SETS AND WORDS

For this section, L is a finite set of letters that we will also use as atoms (see the discussion at the beginning of §4.7). L' is the alphabet L with three extra letters: **{**, **,**, and **}**. It can be shown that there are conservative implementations between **set**(L) and **str**(L'). We prove part of this and sketch the rest because the detailed development calls on several ideas and techniques of elementary set theory, and introducing them here would take us too far afield.

Implementing set(L) in str(L'). The idea is to do formally what we have been doing informally all along, code sets by naming them.

Definition. We define a coding T recursively as follows. If c is an atom, its only T-code is c itself. The only T-code of $\varnothing$, the empty set, is the word **{}**. If $s_1, s_2, \ldots, s_n$ is a sequence of sets and atoms for which T-codes have been defined, and if t_i is any T-code for $s_i (i = 1, 2, \ldots, n)$, then the word **{**$t_1$**,** t_2**,** $\ldots$ **,** t_n**}** is a T-code for the set $\{s_1, s_2, \ldots, s_n\}$.

Exercise 4.8.1. Let **TCODE** be the set of words over L' that are T-codes for members of the domain of **set**(L). Show that **TCODE** is generated over EFS(**str**(L')).

Sets have many T-codes since the order in which members are listed is not important, so the relation **EQUIV**, of set name equivalence, is nontrivial. We give a procedure for it below, after some preliminary items.

The following procedure has as output those words that are nonempty "sequences": $t_1, t_2, \ldots, t_n$ where each t_i is a set name.

NAME_SEQUENCE (1):
TCODE(x)→NAME_SEQUENCE(x);
TCODE(x)→NAME_SEQUENCE(y)→CON(y, **,**, u)
→CON(u, x, v)→NAME_SEQUENCE(v).

The next procedure outputs the relation holding between the words x**,** y and y**,** x, where x and y are set names.

TRANSPOSE_PAIR (2):
TCODE(x)→TCODE(y)→CON(x, **,**, t)→CON(r, y, u)
→CON(y, **,**, s)→CON(s, x, v)→TRANSPOSE_PAIR(u, v).

Next, transposition of adjacent items in a list of set names. There are four cases: the items are at the beginning of the list, at the end, inside, and are the entire list.

TRANSPOSE (2):
NAME_SEQUENCE(v)→TRANSPOSE_PAIR(x, y)→CON(x, **,**, r)
→CON(r, v, m)→CON(y, **,**, s)→CON(s, v, n)
→TRANSPOSE(m, n);
NAME_SEQUENCE(u)→TRANSPOSE_PAIR(x, y)→CON(u, **,**, r)
→CON(r, x, m)→CON(r, y, n)→TRANSPOSE(m, n);
NAME_SEQUENCE(u)→NAME_SEQUENCE(v)
→TRANSPOSE_PAIR(x, y)→CON(u, **,**, r)→CON(r, x, s)
→CON(s, **,**, t)→CON(t, v, m)→CON(r, y, p)→CON(p, **,**, q)
→CON(q, v, n)→TRANSPOSE(m, n);
TRANSPOSE_PAIR(m, n)→TRANSPOSE(m, n).

Exercise 4.8.2. Write a procedure REPEAT whose output is those pairs $\langle m, n \rangle$ where m and n are sequences of set names, and n is like m except that some set name x in the sequence has been replaced by x**,** x.

Exercise 4.8.3. Write a procedure SUBSTITUTE whose output is those quadruples $\langle x, y, m, n \rangle$ where m and n are sequences of set names, and n is like m except that some set name x in the sequence has been replaced by the set name y.

One last minor item. We want a procedure that turns a sequence of set names into a single set name:

SEQ_SET (2):
NAME_SEQUENCE(x)→CON(**{**, x, u)→CON(u, **}**, v)
→SEQ_SET(x, v).

Finally we give the procedure for set name equivalence.

EQUIV (2):
TRANSPOSE(u, v)→SEQ_SET(u, x)→SEQ_SET(v, y)
→EQUIV(x, y);
REPEAT(u, v)→SEQ_SET(u, x)→SEQ_SET(v, y)→EQUIV(x, y);
TCODE(x)→EQUIV(x, x);
TCODE(x)→TCODE(y)→EQUIV(x, y)→EQUIV(y, x);
TCODE(x)→TCODE(y)→TCODE(z)→EQUIV(x, y)
→EQUIV(y, z)→EQUIV(x, z);
EQUIV(x, y)→SUBSTITUTE(x, y, m, n)
→SEQ_SET(m, u)→SEQ_SET(n, v)→EQUIV(u, v).

Exercise 4.8.4. Let **ADDMEM** be the relation on the domain of **set**(L) consisting of triples $\langle x, y, z\rangle$ where z is the set which results when y is added to x as a member. Show that $\mathbf{ADDMEM}^T$ is generated over EFS(**str**(L')).

Thus, T is an implementation of **set**(L) in **str**(L').

Implementing str(L') in set(L). Now we only sketch things. First, as in the previous section, we need representatives for the letters of L', this time in the domain of **set**(L). This is straightforward. Represent each letter of L by itself. And choose any three sets (not atoms) you like as representatives of **{**, **,**, and **}**.

Next, a word is a finite sequence of letters. If we had a set-theoretic counterpart of the notion of finite sequence, we could code a word w by the sequence of representatives for the letters of w. A finite sequence is a function whose domain is a proper initial segment of the nonnegative integers. We have sets that are number representatives (see Chapter 2, §2.4), so all we need is some set-theoretic notion of 1-place function. We can think of a function as a single-valued relation. So what we really need is a set-theoretic counterpart of binary relation. A binary relation is a set of ordered pairs. So, what we ultimately need is a set-theoretic counterpart of an ordered pair. The following is standard: Let $\langle x, y\rangle$ be the set $\{\{x\}, \{x, y\}\}$. It can be shown that this has the essential property one needs: $\langle a, b\rangle = \langle x, y\rangle$ if and only if $a = x$ and $b = y$.

The coding just sketched, call it U, is feasible. It can be shown that U does provide us with an implementation of **str**(L') in **set**(L).

Finally, it can also be shown that both T and U are conservative implementations. Actually, once the work of showing that U is an implementation has been carried out, the additional effort of showing conservativeness is not great. But here we must leave things if we are to avoid writing a book on elementary set theory.

Exercise 4.8.5. Define a function by $J(x, y) = \{\{x\}, \{x, y\}\}$. Show that J is $1-1$. Hint: Suppose that $J(x, y) = J(u, v)$, and show that $x = u$ and $y = v$ by showing

(a) If $x = y$ then $x = u = y = v$.
(b) If $u = v$ then $x = u = y = v$.
(c) If $x \neq y$ and $u \neq v$ then $x = u$ and $y = v$.

Exercise 4.8.6. Show that the graph of J, defined in Exercise 4.8.5, is generated in EFS(**set**(L)).

4.9 IMPLEMENTABILITY

We have seen there are conservative implementations between **integer**, **str**(L) no matter what the size of the alphabet L, and **tree**(**A**) and **set**(**A**), provided the set **A** of atoms is finite. Other data structures can be added to the list. The point is, no matter which of these data structures we pick to work with, it is as if we had *all* of them available. We may as well pick one to concentrate on, and stop talking about the rest. We have chosen EFS(**str**(L)) with an alphabet L of whatever size is handy at the moment. So from now on we essentially return to the settting of Chapter 1, but with the knowledge that the data structure **str**(L) is a very rich one.

It is reasonable to maintain that the *only* data structures that can have real-world computer implementations must have implementations in our technical sense in **str**(L). We can argue intuitively as follows. In order to be able to implement a data structure, it must be possible to describe its members and their properties to a computer or to another person. Descriptions, no matter how exotic, will be in some kind of language. Once one has an accurate description in words, one has the essence of an implementation in **str**(L). We have seen examples of this several times in the present chapter. What we are asserting is not subject to rigorous proof, since we have no abstract notion of computer implementable to work with. But considerations like these lead us to reverse things and make implementability in **str**(L) itself the criterion.

Definition. We say a data structure is *implementable* if it has an implementation in **str**(L) (for some finite alphabet L).

Then in §4.3 we gave an example of an unimplementable data structure: one whose domain consists of all real numbers. Another kind of example was given in §4.4. Since **str**(L) and **integer** have conservative implementations in each other, we could have just as well picked **integer** as our measure of implementability. We did not do so because it is not as handy for our present purposes. But in more theoretical studies of computability, it is often the preferred choice.

Exercise 4.9.1. Invent a data structure you think is appropriate for dealing with signed integers and show that it is implementable.

Exercise 4.9.2. Invent a data structure suitable for work with fractions and show it is implementable. Better yet, show there are *conservative* implementations between it and **integer**.

Exercise 4.9.3. Give two examples of implementable data structures that are not *equivalent* to **str**(L), in the sense that **str**(L) does not have an implementation in them.

4.10 BACKGROUND

The origin of the kind of data structure implementation developed in this chapter is [Gödel 1931]. In his famous paper Gödel showed the incompleteness of a formal system of arithmetic by constructing a self-referential sentence, asserting its own unprovability. In order to make statements about numbers act as if they were statements about statements, Gödel introduced a sytem of coding words by numbers that turned syntax manipulations into arithmetic operations. Today, such codings are often known as Gödel numberings. We considered a Gödel numbering in §4.6, though not the one Gödel himself used.

The coding of words by trees used in §4.7 is essentially the way lists are represented, using trees, in LISP, while the representatives of trees as words are essentially the S-expressions of LISP. Both notions can be found in [McCarthy 1960]. The detailed language description itself may be found first in [McCarthy et al. 1962].

The way of representing words as sets sketched in §4.8 is standard in set theory. It is nothing more than the conventional set theoretic notion of finite sequence. The other direction simply takes the common method of naming sets and makes it into a coding. [Montague 1968] seems to have first taken seriously the idea that working with finite sets of finite sets of . . . of finite sets of atoms could give a useful notion of computation.

If one takes the family of all data structures as objects, and uses implementations as morphisms, one has a natural category. It, and generalizations of it, were studied in [Fitting 1981].

REFERENCES

[Fitting 1981] *Fundamentals of Generalized Recursion Theory,* M. Fitting, North Holland Publishing Co., Amsterdam, 1981.

[Gödel 1931] Über formal unentscheidbare Sätze der Principia mathematica und verwandter Systeme I, K. Gödel, *Montshefte für Mathematik und Physik,* Vol. 38, pp. 173–198; English translations in *The Undecidable,* M. Davis, editor, Raven Press, Hewlett, N. Y., pp. 4–38, 1965, and in *From Frege to Gödel,* J. van Heijenoort, editor, Harvard University Press, Cambridge, Mass., pp. 592–616, 1967.

[McCarthy 1960] Recursive functions of symbolic expressions and their computation by machine, J. McCarthy, *Communications of the Association for Computing Machinery,* Vol. 3, pp. 184–195, 1960.

[McCarthy et al. 1962] *The Lisp 1.5 Programmer's Manual,* J. McCarthy, P. Abrahams, D. Edwards, T. Hart, and M. Levin, MIT Press, Cambridge, Mass., 1962.

[Montague 1968] Recursion theory as a branch of model theory, R. Montague, in *Logic, Methodology, and Philosophy of Science III, Proc. of the 1967 Congress,* B. van Rootselaar et al., editors, North Holland Publishing Co., Amsterdam, pp. 63–86, 1968.

5

THE CHURCH–TURING THESIS

5.1 INTRODUCTION

By now you should be familiar with the language EFS(**str**(L)), its style and, from Chapter 4, its richness. But it is a relatively unfamiliar kind of language. It is time to consider implementation issues, and the relationship between EFS(**str**(L)) and more conventional languages.

We present a mathematical model of a computer, called a *register machine* and, eventually, we show that EFS(**str**(L)) can be implemented on such a machine. In the process of doing so we introduce two more programming languages, IMP, which is a deterministic Pascal-like language and LOG, based on the logical notation of Chapter 2, and we show they are all equivalent in computing power. This is taken as an argument for the *Church–Turing Thesis,* which says that there is an absolute notion of computation that is independent of machine or language.

It should be emphasized that *efficiency* issues play no role here. We show *some* implementation of EFS(**str**(L)) on register machines is possible; in practice, ours would be extraordinarily time consuming. Much current research in logic programming is devoted to the efficient implementation of languages related to EFS(**str**(L)).

5.2 REGISTER MACHINES

Definition. A *register machine* for the alphabet L consists of an unlimited set $R_1, R_2, \ldots$ of *registers,* each capable of *storing* exactly one word over $L,$ and the following *instruction set*:

1. Zi (zero a register) Store the empty word, Δ, in register R_i;

2. For each letter a of the alphabet $L,$
   $S_a i$ (successor) Replace the word stored in R_i by the result of adjoining the letter a to the (right) end;

3. Fi (first) Replace the word stored in R_i by its first letter provided the word is not Δ, and by Δ otherwise;

4. Bi (butfirst) Replace the word stored in R_i by the result of deleting its first letter provided the word is not Δ, and by Δ otherwise;

5. Ci, j	(copy)	Store in R_i a copy of the word stored in R_j;
6. Ji, j, k	(jump)	If R_i and R_j are storing the same word jump to program instruction k;
7. Stop	(halt)	End of program execution.

For example, if R_3 contains the word *abcb*, and the instruction S_c3 is executed, then R_3 will contain *abcbc*. Then if $B3$ is executed, R_3 will contain *bcbc*. Finally, if $F3$ is executed, R_3 will contain *b*. Instruction 2, successor, is really a finite family of instructions, one for each member of the alphabet.

There are several places where idealizations of real computers are made when working with register machines. There will be no bound on program length or running time. There is no bound on the length of a word that can be stored in a single register, and there is no bound on the number of registers. This last idealization is not significant, however. Each particular program will only use a fixed, finite number of registers, a number which can not change during program execution. And it can be shown that there is a number, N, such that *any* programmable function can be computed without using more than N registers (see Exercise 6.6.2). But since a word can be used to "remember" information, placing no restriction on word length is what really says we are assuming unlimited storage.

Definition. A *program* for a register machine is a finite, numbered sequence of instructions such that, if Ji, j, k occurs in the program, the program has at least k instructions; and the last instruction is Stop.

Because of the restriction on jumps, all jumps in a program must be meaningful. Because the last instruction must be a halt instruction, we cannot "run off the end" of a program. Finally, we will assume that all registers otherwise uninitialized contain Δ, so successor instructions and the like are always applicable. It follows that in executing a program we will never find ourselves in an undefined situation. We will think of a program as instructions for computing a function. Certain *input registers* are filled, then the program is executed and if it halts, the contents of a designated *output register* is taken to be the value of the function for the input supplied. If the program never halts, the function does not have that input in its domain. To make things simple, we will use $R_1, \ldots, R_n$ as input registers when computing a function of n arguments, and we will reuse R_1 as an output register. Then the only thing we need to specify besides the program itself is *how many* inputs there will be.

Definition. A *module* is a pair consisting of a program and a number, called the *input specification number*, intended to record the number of inputs.

In the examples below, L can be any nonempty alphabet. Since it will be convenient to have number representatives available, we choose one letter of L

for this purpose, and use the symbol 1 for it. We think of a string of n 1's as representing the number n. We continue using Δ to symbolize the empty word, but we also think of it as representing the number 0. Finally, we sometimes think of 1 as representing *truth* and Δ as representing *falsehood*. Understand, these are all just words over L, but it is convenient from time to time to attach these meanings to them.

The first example is a module we call LENGTH. The input specification number is 1, and the output will be (the number representative for) the length of the input. The program part of the module is:

1	$Z2$	; R_2 remembers input length counted so far.
2	$Z3$	; R_3 holds Δ for comparison purposes.
3	$J1, 3, 7$	; if input not yet erased,
4	$B1$	; shorten input by one letter and
5	S_12	; add one to R_2.
6	$J3, 3, 3$	; Always jump.
7	$C1, 2$	; Put output in R_1.
8	Stop	

We have included comments as informal aids to understanding. They are not officially part of the program. Since all registers except those used as input are assumed to contain Δ at the start of program execution, instructions 1 and 2 are not strictly necessary, but they help make the program clearer. In fact, we will initialize every register we intend to use. Instruction 6 is a conditional jump that must always be taken; in effect, it is a GOTO. This is an example of the simulation of instructions not in the original set. We will see much more of this. The program only needs three registers, the three specifically referred to by its program instructions.

We have not formally said what it means to *run* a program, we will do that in the next section. But an informal understanding should suffice for now. We suggest that you trace the steps of execution for the examples in this section, for several choices of inputs.

The second example is a module we call SHORTER. The input specification number is 2. The output is *true* (1) if input 1 (placed in R_1) is shorter than input 2 (placed in R_2) and is *false* (Δ) otherwise. The program part of the module is:

1	$Z3$	; R_3 holds "the answer", initialized to false.
2	$Z4$	; R_4 holds Δ for comparison purposes.
3	$J2, 4, 9$	; Answer should be false.
4	$J1, 4, 8$	; Answer should be true.
5	$B1$	; Shorten input 1
6	$B2$	; and input 2.
7	$J4, 4, 3$	; Always jump.
8	S_13	; Change answer to true.
9	$C1, 3$	; Put output in R_1.
10	Stop	

Exercise 5.2.1. Modify the SHORTER module so that the output is the shorter word itself. If both are the same length, output the first of the two inputs.

Exercise 5.2.2. Write a number-checking module. It should take a single input, and output true if the input word consists of all 1's, and output false otherwise.

Exercise 5.2.3.

(a) Write an addition module. When given as input two number representatives, it produces as output the representative of their sum. (When supplied with nonnumber representatives, the output can be whatever you want.)

(b) Write a similar module for multiplication.

Exercise 5.2.4. Suppose we added to the instruction set a concatenation instruction, CON i, j, meaning replace the contents of R_i with its former contents concatenated with the contents of R_j. Show that the first-letter-of instruction, F, becomes redundant.

5.3 COMPUTATIONS

We make the notion of running a register machine module precise by defining computation trace.

Definition. Suppose that we have a module M for the language L, with input specification number n, and with a program P with m instructions, which uses registers $R_1, \ldots, R_r$. A *state vector* (or just *state*) for module M is an $n+r+2$ tuple $\langle i, n, a_1, \ldots, a_n, w_1, \ldots, w_r \rangle$, where i is an integer, $1 \leq i \leq m$, and $a_1, \ldots, a_n, w_1, \ldots, w_r$ are words over L.

The state vector $\langle i, n, a_1, \ldots, a_n, w_1, \ldots, w_r \rangle$ is intended to capture the following information: We are running a module that was given the n-tuple $a_1, \ldots, a_n$ as input; we are about to execute program instruction number i; and the registers $R_1, \ldots, R_r$ currently store the words $w_1, \ldots, w_r$ respectively. This should account for the definition of trace below.

Definition. Suppose that we have a module M for the language L, with input specification number n, and with a program P with m instructions, which uses registers $R_1, \ldots, R_r$. By a *trace* for M using $a_1, \ldots, a_n$ as input we mean a sequence (not necessarily finite) of state vectors for M meeting the following conditions.

0. The first state in the sequence is $\langle 1, n, a_1, \ldots, a_n, a_1, \ldots, a_n, \Delta, \ldots, \Delta \rangle$.

Next, suppose that $\langle p, n, a_1, \ldots, a_n, w_1, \ldots, w_r \rangle$ is a state in the trace; we say what the next state must be. There are several cases depending on what the

pth program instruction of P is.

1. If instruction p is Zi, the next state is
 $\langle p+1, n, a_1, \ldots, a_n, w_1, \ldots, w_{i-1}, \Delta, w_{i+1}, \ldots, w_r \rangle$;
2. If instruction p is $S_a i$, the next state is
 $\langle p+1, n, a_1, \ldots, a_n, w_1, \ldots, w_{i-1}, w_i', w_{i+1}, \ldots, w_r \rangle$
 where w_i' is the result of concatenating w_i and a.
3. If instruction p is Fi, the next state is
 $\langle p+1, n, a_1, \ldots, a_n, w_1, \ldots, w_{i-1}, w_i', w_{i+1}, \ldots, w_r \rangle$
 where w_i' is the first letter of w_i if w_i is not Δ, and otherwise w_i' is Δ.
4. If instruction p is Bi, the next state is
 $\langle p+1, n, a_1, \ldots, a_n, w_1, \ldots, w_{i-1}, w_i', w_{i+1}, \ldots, w_r \rangle$
 where w_i' is w_i without its first letter if w_i is not Δ, and otherwise w_i' is Δ.
5. If instruction p is Ci, j, the next state is
 $\langle p+1, n, a_1, \ldots, a_n, w_1, \ldots, w_{i-1}, w_j, w_{i+1}, \ldots, w_r \rangle$;
6. If instruction p is Ji, j, k, the next state is
 $\langle k, n, a_1, \ldots, a_n, w_1, \ldots, w_r \rangle$ if $w_i = w_j$ and is
 $\langle p+1, n, a_1, \ldots, a_n, w_1, \ldots, w_r \rangle$ otherwise.
7. If instruction p is Stop, there is *no* next state.

Definition. If a trace is finite, that is, if it has a last state, then the module is said to *halt* on the given inputs. If so, the value of R_1 in the final state is the *output*.

For example, the following is a trace of the LENGTH module from §5.2, using input *ab*.

$\langle 1, 1, ab, ab, \Delta, \Delta \rangle$
$\langle 2, 1, ab, ab, \Delta, \Delta \rangle$
$\langle 3, 1, ab, ab, \Delta, \Delta \rangle$
$\langle 4, 1, ab, ab, \Delta, \Delta \rangle$
$\langle 5, 1, ab, b, \Delta, \Delta \rangle$
$\langle 6, 1, ab, b, 1, \Delta \rangle$
$\langle 3, 1, ab, b, 1, \Delta \rangle$
$\langle 4, 1, ab, b, 1, \Delta \rangle$
$\langle 5, 1, ab, \Delta, 1, \Delta \rangle$
$\langle 6, 1, ab, \Delta, 11, \Delta \rangle$
$\langle 3, 1, ab, \Delta, 11, \Delta \rangle$
$\langle 7, 1, ab, \Delta, 11, \Delta \rangle$
$\langle 8, 1, ab, 11, 11, \Delta \rangle$

In the final state the value of R_1 is 11, which is thus the output for an input of *ab*.

Definition. Module M, consisting of input specification number n and program P, *computes the function* f if: (1) the domain of f consists of those n-tuples on which P halts when used as input, and (2) for $\langle w_1, \ldots, w_n \rangle$ in the domain of f, the output of M, on $\langle w_1, \ldots, w_n \rangle$ as input, is $f(w_1, \ldots, w_n)$. We call f *register machine computable* if some module computes it.

Exercise 5.3.1. Write out traces for the SHORTER module from §5.2, using various inputs.

Exercise 5.3.2. Give an example of a module that, for some input, produces no output. That is, it never halts. Write out several steps of a trace.

Probably you did the exercise above in a way that produced a periodic trace. Consider an analogy with infinite decimals. The periodic ones constitute a relatively small class (they correspond to rational numbers). There are many ways a decimal can be infinite without presenting any discernible pattern. Similarly, it is possible for a trace to be infinite but in such a complicated way that we would not have guessed it by looking at the module and the input. We will return to this, rather vague, point in the next chapter.

5.4 COMPUTATION STATES

Definition. Suppose that we have a module M for the alphabet L. Any state vector that can occur in a trace for M using an arbitrary input is called a *computation state* for M.

We characterize the set of computation states for a given module as the least fixed point of a monotone operator. This is in preparation for the writing of an EFS procedure in §5.5. The operator is simply one which, when supplied with a set of computation states for M, produces the set of "next" computation states.

Definition. Let M be a module over the alphabet L, with input specification number n, and with a program P, which uses registers $R_1, \ldots, R_r$. By Φ_M we mean the operator taking sets of state vectors for M to sets of state vectors for M, as follows. $\Phi_M(S)$ is the smallest set of state vectors for M meeting the conditions:

0. $\langle 1, n, a_1, \ldots, a_n, a_1, \ldots, a_n, \Delta, \ldots, \Delta \rangle \in \Phi_M(S)$ for every choice $a_1, \ldots, a_n$ of words from L^*;

and if $\langle p, n, a_1, \ldots, a_n, w_1, \ldots, w_r \rangle \in S$ then

1. If instruction p is of P is Zi, then $\langle p+1, n, a_1, \ldots, a_n, w_1, \ldots, w_{i-1}, \Delta, w_{i+1}, \ldots, w_r \rangle \in \Phi_M(S)$.
2. If instruction p of P is $S_a i$, then $\langle p+1, n, a_1, \ldots, a_n, w_1, \ldots, w_{i-1}, w_i', w_{i+1}, \ldots, w_r \rangle \in \Phi_M(S)$ where w_i' is the result of concatenating w_i and a.

3. If instruction p of P is Fi, then
   $\langle p+1, n, a_1, \ldots, a_n, w_1, \ldots, w_{i-1}, w_i', w_{i+1}, \ldots, w_r\rangle \in \Phi_M(S)$
   where w_i' is the first letter of w_i if w_i is not Δ, and otherwise w_i' is Δ.
4. If instruction p of P is Bi, then
   $\langle p+1, n, a_1, \ldots, a_n, w_1, \ldots, w_{i-1}, w_i', w_{i+1}, \ldots, w_r\rangle \in \Phi_M(S)$
   where w_i' is w_i without its first letter if w_i is not Δ, and otherwise w_i' is Δ.
5. If instruction p of P is Ci, j, then
   $\langle p+1, n, a_1, \ldots, a_n, w_1, \ldots, w_{i-1}, w_j, w_{i+1}, \ldots, w_r\rangle \in \Phi_M(S)$.
6. If instruction p of P is Ji, j, k, then
   $\langle k, n, a_1, \ldots, a_n, w_1, \ldots, w_r\rangle \in \Phi_M(S)$ if $w_i = w_j$ and
   $\langle p+1, n, a_1, \ldots, a_n, w_1, \ldots, w_r\rangle \in \Phi_M(S)$ otherwise.

This operator is easily seen to be monotone. It is of interest to us because of the following.

Proposition 5.4.1. *Let* **F** *be the smallest fixed point of the operator* Φ_M, *for a given module* M. *Then* **F** *is exactly the set of computation states for* M.

The proof of this proposition breaks into two parts, which are left as exercises.

Exercise 5.4.1. Let **T** be *any* (semi) fixed point of Φ_M. Show that if $\langle p, n, a_1, \ldots, a_n, w_1, \ldots, w_r\rangle$ is a computation state for M, then $\langle p, n, a_1, \ldots, a_n, w_1, \ldots, w_r\rangle \in \mathbf{T}$. *Hint*: If $\langle p, n, a_1, \ldots, a_n, w_1, \ldots, w_r\rangle$ is a computation state for M, it occurs as some line in a trace for M, say as line q. Use induction on q.

Exercise 5.4.2. Let **F** be the *least* fixed point of Φ_M. Show that every member of **F** is a computation state for M. *Hint*: Let **T** be the collection of all computation states for M. Show that **T** is a semifixed point of Φ_M and use generalized induction, Theorem 3.3.3.

We allow register machine programs to have many halt instructions. But all except the final one can be replaced by GOTO's to that one without essentially affecting program behavior. (We saw in §5.2 how to simulate GOTO's.) Thus, the following corollary is fully general. It says that the graph of the function that module M computes can be recovered from the least fixed point of Φ_M.

Corollary 5.4.2. *Suppose that* M *is a module over alphabet* L, *with input specification number* n *and program* P. *Suppose further that* P *contains a single halt instruction, say as instruction number* s. *Then* M *with input* $a_1, \ldots, a_n$ *has output* b *if and only if, for some words* $w_2, \ldots, w_r$, $\langle s, n, a_1, \ldots, a_n, b, w_2, \ldots, w_r\rangle$ *is in the least fixed point of* Φ_M.

Proof. By definition, M with $\langle a_1, \ldots, a_n\rangle$ as input gives b as output if the trace for M with $\langle a_1, \ldots, a_n\rangle$ as input has a final state and, in that final state b is the value of R_1. Since the only halt instruction is the sth, that final state must be $\langle s, n, a_1, \ldots, a_n, b, w_2, \ldots, w_r \rangle$. Also, if $\langle s, n, a_1, \ldots, a_n, b, w_2, \ldots, w_r \rangle$ occurs as a state in a trace for M, it must be the final state because instruction s is a halt instruction. The result then follows from Proposition 5.4.1.

5.5 TRANSLATING REGISTER MACHINES INTO EFS(str(L))

We show that for any register machine module M over the alphabet L, there is a procedure of EFS(**str**(L)) whose output is the graph of the function that M computes. Thus, EFS procedures provide a notion of computation that is at least as strong as that provided by register machines. We begin by showing the operator Φ_M, from the previous section, corresponds to an EFS procedure. A minor modification first, however. As defined in §5.3, state vectors were tuples of words over L and *numbers*. From now on, instead of using numbers, we use their *number representatives* in L^*. For a number n, the symbol $\bar{n}$ stands for a string of n 1's, where 1 is a designated letter of L. For example, $\bar{3} = 111$. This makes $\bar{n}$ a word over L. We never use $\bar{x}$ where x is a variable, for this has no meaning; we only use $\bar{n}$ where n is a particular number. From now on, state vectors involve number representatives, not numbers.

Theorem 5.5.1. *Let* M *be a module over the alphabet* L. *The operator* Φ_M *is defined by a procedure of EFS*(**str**(L)).

Proof. Say that M has input specification number n, and program P using registers $R_1, \ldots, R_r$. The operator Φ_M is defined by the EFS(**str**(L)) procedure with header

PHI_OUT $(n + r + 2)$ INPUT PHI_IN$(n + r + 2)$:

and with a body determined by the following rules.

0. PHI_OUT$(\bar{1}, \bar{n}, x_1, \ldots, x_n, x_1, \ldots, x_n, \Delta, \ldots, \Delta)$ is a procedure statement.
1. If instruction p of P is Zi, then

   PHI_IN$(\bar{p}, \bar{n}, x_1, \ldots, x_n, y_1, \ldots, y_r)$
   $\rightarrow$ PHI_OUT$(\overline{p+1}, \bar{n}, x_1, \ldots, x_n, y_1, \ldots, y_{i-1}, \Delta, y_{i+1}, \ldots, y_r)$

   is a procedure statement.
2. If instruction p of P is $S_a i$, then

   PHI_IN$(\bar{p}, \bar{n}, x_1, \ldots, x_n, y_1, \ldots, y_r)$
   $\rightarrow$ CON(y_i, a, z)
   $\rightarrow$ PHI_OUT$(\overline{p+1}, n, x_1, \ldots, x_n, y_1, \ldots, y_{i-1}, z, y_{i+1}, \ldots, y_r)$

   is a procedure statement.
3. If instruction p of P is Fi, then

   PHI_IN$(\bar{p}, \bar{n}, x_1, \ldots, x_n, y_1, \ldots, y_r)$
   $\rightarrow$ LETTER(z) $\rightarrow$ CON(z, w, y_i)
   $\rightarrow$ PHI_OUT$(\overline{p+1}, \bar{n}, x_1, \ldots, x_n, y_1, \ldots, y_{i-1}, z, y_{i+1}, \ldots, y_r)$

and

$$\text{PHI\_IN}(\bar{p}, \bar{n}, x_1, \ldots, x_n, y_1, \ldots, y_{i-1}, \Delta, y_{i+1}, \ldots, y_r)$$
$$\rightarrow \text{PHI\_OUT}(\overline{p+1}, \bar{n}, x_1, \ldots, x_n, y_1, \ldots, y_{i-1}, \Delta, y_{i+1}, \ldots, y_r)$$

are procedure statements.

4. If instruction p of P is Bi, then

$$\text{PHI\_IN}(\bar{p}, \bar{n}, x_1, \ldots, x_n, y_1, \ldots, y_r)$$
$$\rightarrow \text{LETTER}(z) \rightarrow \text{CON}(z, w, y_i)$$
$$\rightarrow \text{PHI\_OUT}(\overline{p+1}, \bar{n}, x_1, \ldots, x_n, y_1, \ldots, y_{i-1}, w, y_{i+1}, \ldots, y_r)$$

and

$$\text{PHI\_IN}(\bar{p}, \bar{n}, x_1, \ldots, x_n, y_1, \ldots, y_{i-1}, \Delta, y_{i+1}, \ldots, y_r)$$
$$\rightarrow \text{PHI\_OUT}(\overline{p+1}, \bar{n}, x_1, \ldots, x_n, y_1, \ldots, y_{i-1}, \Delta, y_{i+1}, \ldots, y_r)$$

are procedure statements.

5. If instruction p of P is Ci, j then

$$\text{PHI\_IN}(\bar{p}, \bar{n}, x_1, \ldots, x_n, y_1, \ldots, y_r)$$
$$\rightarrow \text{PHI\_OUT}(\overline{p+1}, \bar{n}, x_1, \ldots, x_n, y_1, \ldots, y_{i-1}, y_j, y_{i+1}, \ldots, y_r)$$

is a procedure statement.

6. If instruction p of P is Ji, j, k then

$$\text{PHI\_IN}(\bar{p}, \bar{n}, x_1, \ldots, x_n, y_1, \ldots, y_r)$$
$$\rightarrow \text{EQUAL}(y_i, y_j)$$
$$\rightarrow \text{PHI\_OUT}(\bar{k}, \bar{n}, x_1, \ldots, x_n, y_1, \ldots, y_r)$$

and

$$\text{PHI\_IN}(\bar{p}, \bar{n}, x_1, \ldots, x_n, y_1, \ldots, y_r)$$
$$\rightarrow \text{NOT\_EQUAL}(y_i, y_j)$$
$$\rightarrow \text{PHI\_OUT}(\overline{p+1}, \bar{n}, x_1, \ldots, x_n, y_1, \ldots, y_r)$$

are procedure statements (Exercise 1.5.3 asked you to program EQUAL and NOT_EQUAL).

This completes the description of the procedure PHI_OUT. It is easy to see that PHI_OUT defines the operator Φ_M, since we have exactly paralleled the definition from §5.4. This ends the proof.

Corollary 5.5.2. *If a function is register machine computable over an alphabet* L, *then the graph of that function is generated by an EFS*(**str**(L)) *procedure.*

Proof. Suppose the n-ary function f is computed by the module M. Without loss of generality we can assume that the program part of M has only a single halt statement, say instruction s. By the preceding theorem, the operator Φ_M is defined by an EFS(**str**(L)) procedure. This operator has a smallest fixed point, call it **F**. By the First Recursion Theorem, **F** is generated by an EFS(**str**(L)) procedure. By Corollary 5.4.2, the graph of the function f is just

$$\{\langle a_1, \ldots, a_n, b\rangle \mid (\exists w_2) \ldots (\exists w_r)\langle \bar{s}, \bar{n}, a_1, \ldots, a_n, b, w_2, \ldots, w_r\rangle \in \mathbf{F}\}.$$

Since generated relations are closed under sections and existential quantification, the graph of f is generated.

Exercise 5.5.1. Follow the method given above, start with the LENGTH module from §5.2, and produce an EFS(**str**(L)) procedure whose output is the graph of the function computed by LENGTH.

Exercise 5.5.2. Turn the module SHORTER, also from §5.2, into an EFS(**str**(L)) procedure.

5.6 AN IMPERATIVE LANGUAGE IMP

We have shown how to translate register machine mdoules into EFS procedures. Such a translation is neither difficult nor surprising given the obvious power of EFS. To go the other way, to implement EFS(**str**(L)) procedures on register machines, is more work. One problem is the weakness of the language used to program register machines. In this section we introduce a Pascal-like language, IMP, that is easier to use. IMP is a fully *deterministic* language. Unlike EFS, but like register machines, in running an IMP program there is a command that must be executed first, a command that must be executed next, and so on. And these are really commands, orders to do something. Languages like IMP are sometimes called *imperative* languages, hence the name IMP. Although an operational and a denotational semantics for IMP can be given, it would take us too far afield. Instead, we first explain informally how IMP programs are supposed to behave. Then, we show how each IMP program can be turned into a corresponding register machine program. We take this translation into register machine language as the "official" characterization of IMP program behavior. Translatability is what we are primarily interested in. IMP provides us with a more easily understood way of giving instructions to register machines.

For the following definitions, L is some nonempty alphabet. We continue to make the assumption that L contains no letter that could be confused with a symbol having a special meaning. As usual, this allows us to omit opening and closing quotes when writing words over L. Since the alphabet L is otherwise arbitrary we are really defining a family of languages, IMP(L), indexed by L.

We want to have counting loops. We choose one letter from the alphabet L, denote it by 1, and use a string of n 1's to represent the number n, denoting it by $\bar{n}$. Then $\bar{1} = 1$ and $\bar{0} = \Delta$.

The *identifiers* of IMP(L) are the same as in EFS languages. The *constants* of IMP(L) are words over L. We need some word-manipulating operations. There are a large number of possibilities that will produce languages of equivalent power. As we did with register machines, we have chosen a mechanism for building words up, and a mechanism for tearing them apart. To build words up we have *concatenation,* treated as a binary function, rather than as a relation. To tear words apart, we have the BUTFIRST function (taken from LISP via

LOGO). Finally, we have a LENGTH function to turn arbitrary words into number representatives.

Definition. The *terms* of IMP(L) are identifiers, constants, and expressions of the form $t + u$, BUTFIRST(t) and LENGTH(t), where t and u are identifiers or constants. The intention is, the value of $t + u$ is to be the value of t concatenated with the value of u. The value of BUTFIRST(t) is to be the value of t without its first letter provided the value of t is not Δ, and is to be Δ otherwise. The value of LENGTH(t) is to be the string of 1's of the same length as the value of t.

The definition of term does not compound. Thus, BUTFIRST($t + u$) is not a term, for instance. Also, concatenation is strictly binary, so $t + u + v$ is not a term. (But see Exercise 5.8.1). For numbrs n and m, $\bar{n} + \bar{m} = \overline{n + m}$, where the $+$ on the left denotes concatenation, whereas that on the right denotes addition.

Definition. *Conditions* of IMP(L) are any expressions of the forms

$t = u$ or $t < u$

where t and u are identifiers or constants. The intention is, if t and u have exactly the same value, $t = u$ is *true*, otherwise $t = u$ is *false*. If the value of t is a shorter word than the value of u, $t < u$ is *true*, otherwise $t < u$ is *false*.

Just as with terms, conditions do not compound. This is to simplify translation into register machine programs. But see Exercise 5.8.2 for the implementation of Boolean combinations of conditions.

Definition. *Assignment statements* of IMP(L) are expressions of the form

IDENTIFIER: $= t$

where IDENTIFIER is any identifier and t is a term. The intention is that the new value of IDENTIFIER should be the current value of t.

Statements are now built up recursively from assignment statements much like the logical expressions of Chapter 2 were built up from atomic ones.

Definition of IMP(L) *statement*

1. Any IMP(L) assignment statement is an IMP(L) statement.
2. If $S_1, S_2, \ldots, S_n$ is a sequence of IMP(L) statements then

   BEGIN $S_1; S_2; \ldots; S_n$ END

   is an IMP(L) statement. The intention is, to execute this statement, first execute S_1, then execute $S_2, \ldots,$ and finally execute S_n.

3. If C is a condition of IMP(L) and S_1 and S_2 are IMP(L) statements then

   IF C THEN S_1 ELSE S_2

   is an IMP(L) statement. The intention is, if C is true then execute S_1, otherwise execute S_2.

4. If C is a condition of IMP(L) and S is an IMP(L) statement then

   WHILE C DO S

   is an IMP(L) statement. The intention is, to execute WHILE C DO S, if C is false there is nothing to do, and if C is true execute S, then execute WHILE C DO S again.

5. If t and u are identifiers or constants, N is an identifier, and S is an IMP(L) statement then

   FOR $N := t$ TO u DO S

   is an IMP(L) statement. N is the *loop counter* and S is the *loop body*. The intention is, to execute FOR $N := t$ TO u DO S where t is an identifier, if the length of the value of t is greater than the length of the value of u there is nothing to do, and otherwise execute the assignment $N := t$, execute S, execute $t := t + \bar{1}$, then execute FOR $N := t$ TO u DO S. If t is a constant, to execute FOR $N := t$ TO u DO S, execute $X := t$, then FOR $N := X$ TO u DO S, where X is an otherwise unused identifier.

In counting loops the loop bounds are effectively identified with number representatives; nothing about their values matters except lengths. In presenting IMP(L) statements we will freely use spaces and new lines to enhance readability. These are not official parts of the language, however.

Exercise 5.6.1. Show that the following is an IMP(L) statement.

```
IF X_NUM < Y_NUM THEN DIFF: = 0̄
ELSE
    BEGIN
    DIFF: = X_NUM;
    FOR N: = 1̄ TO Y_NUM DO
        DIFF: = BUTFIRST(DIFF)
    END
```

Programming languages must provide some interaction with the outside world. Since we are only interested in computational aspects, we assume that input has been supplied before we start running a program, and output is taken care of after the program halts. We set aside certain identifiers to represent inputs. Before a program starts execution they are given values that cannot be changed thereafter. Another identifier is set aside to represent output. When a program ceases execution (if ever), the final value of the output identifier is considered to be the output. We record the necessary input/output information in a program header, as we did with EFS.

Definition. A *program header* is an expression of the form

OUTPUT (IN_ONE, IN_TWO, . . . , IN_LAST):

where OUTPUT, IN_ONE, IN_TWO, . . . , IN_–LAST are identifiers. IN_ONE, IN_TWO, . . . , IN_LAST are said to be *reserved* by this program header. The intention is, they will represent input, and OUTPUT will represent output.

Definition. An IMP(L) *program* consists of a *program header* followed by a *program body* made up of a single IMP(L) statement, ending with a period. In the program body, no identifier that is reserved by the program header can appear on the left side of an assignment statement, and in any counting loop that occurs as part of the program body the loop counter may not be reserved and may not appear on the left side of an assignment statement within the loop body.

The restriction on reserved identifiers ensures that a program can not change its own input; programs of IMP(L) have output but no side effects. The restriction on loop counters guarantees that they can be used only as counters, in well-behaved, structured ways.

For example, the following is an IMP(L) program (see Exercise 5.6.1).

```
DIFF (X_NUM, Y_NUM):
    IF X_NUM < Y_NUM THEN DIFF: = 0̄
    ELSE
        BEGIN
        DIFF: = X_NUM;
        FOR N: = 1̄ TO Y_NUM DO
            DIFF: = BUTFIRST(DIFF)
        END.
```

You should verify informally that executing this program, having previously assigned number representative values to X_NUM and Y_NUM, causes the output identifier DIFF to have as its value the number representative of length of X_NUM – length of Y_NUM, or 0 if the result of the subtraction is negative.

Exercise 5.6.2. Write a number-checking program. It should take a single input, and output 1 if the input word consists of all 1's, and output 0 otherwise.

Exercise 5.6.3. Write an addition program. When given as input two number representatives, it produces as output the representative of their sum. (When supplied with nonnumber representatives, the output can be whatever you wish.)

Exercise 5.6.4. Write a multiplication program.

Exercise 5.6.5. We did not include a FIRST function to go with BUTFIRST. Show it is not necessary. That is, write a program such that, if the input is Δ the output is Δ, and otherwise the output is the first letter of the input.

5.7 IMPLEMENTING IMP

We show how to translate IMP programs into register machine modules. This translation can be taken as a *definition* of the behavior of IMP programs. After this, we needn't work directly with register machines any more.

Suppose that S is a *statement* of IMP(L). We define a translation, v, turning S into a register machine program over L. To begin, assign to each identifier of S, say IDENTIFIER, a distinct register, v(IDENTIFIER). In addition to these, we may need other registers for "scratch work." There is a certain amount of arbitrariness in our choice of which registers to use for which purposes, but this makes no essential difference.

S, being an IMP(L) statement, is built up from assignment statements using structured programming constructs as defined in §5.6. We define the translation v by beginning with assignment statements that occur in S, and working upward, paralleling the construction of S itself.

Suppose that the assignment statement IDENTIFIER: = t occurs in S. There are several cases depending on the term t.

Assignment statement case 1: t is an identifier, say OTHER. Then the associated register machine program, v(IDENTIFIER: = OTHER), is just

1 Cv(IDENTIFIER), v(OTHER)

Assignment statement case 2: t is a constant, say, for example, bcb. Then v(IDENTIFIER: = bcb) is the register machine program

1 Zv(IDENTIFIER)
2 S_bv(IDENTIFIER)
3 S_cv(IDENTIFIER)
4 S_bv(IDENTIFIER)

Assignment statement case 3: t is ONE + TWO where ONE and TWO are identifiers. This is the most complicated of the assignment statement cases. To be specific, say the alphabet L is $\{b, c\}$. Let R_i, R_j, R_k, R_l, and R_m be unused registers, not associated with any identifier of S. We use these for "scratch work" as follows. R_i holds a copy of the value of TWO, which we can take apart without affecting v(TWO) (program inputs are not to be modified during program execution). R_j contains the constant Δ, to be used for comparison purposes. Likewise, R_k and R_l hold the one-letter words b and c. R_m successively contains single letters taken from the value of TWO. Now, v(IDENTIFIER: = ONE +

TWO) is the register machine program

1	Cv(IDENTIFIER), v(ONE)	
2	Ci, v(TWO)	; copy of second input
3	Zj	; R_j is Δ reference
4	Zk	
5	$S_b k$	; R_k is b reference
6	Zl	
7	$S_c l$	; R_l is c reference
8	Ji, j, 17	; test for finish of work
9	Cm, i	
10	Fm	; R_m is first letter of R_i
11	Bi	; shorten R_i
12	Jm, k, 15	; test for b
13	$S_c v$(IDENTIFIER)	; else must be c
14	Jj, j, 8	; GOTO 8
15	$S_b v$(IDENTIFIER)	; b case
16	Jj, j, 8	; GOTO 8
17	Stop	

Exercise 5.7.1. There are more assignment statement cases. Identify them and show how to define v for them.

Now we define v on more complicated parts of S, finally working up to S itself. In what follows, S_1 and S_2 are substatements of S.

Sequence case: Suppose that v has been defined on the statements S_1 and S_2; we define v on BEGIN S_1; S_2 END. The general case of an arbitrary sequence of statements is handled in a similar way. We can assume that $v(S_1)$ has only a single Stop instruction, as the last line; any others could be replaced by jumps to that last line. Say the last line of $v(S_1)$ is number n with instruction Stop. Delete it from $v(S_1)$, add $n-1$ to every line number of $v(S_2)$, including those occurring in jump instructions, and append the revised $v(S_2)$ to the end of the revised $v(S_1)$ to get v(BEGIN S_1; S_2 END).

While loop case 1: Suppose that v has been defined on the statement S_1, and ONE and TWO are identifiers. We define v on WHILE ONE = TWO DO S_1. Again, we assume that $v(S_1)$ has a single Stop instruction. First move $v(S_1)$ down to make room for a new first line by adding 1 to each line number of $v(S_1)$, including those occurring in jump instructions. Say the last line of $v(S_1)$ is now number n; its instruction is Stop. Replace this with a new line n containing the instruction Jv(ONE), v(ONE), 1 (a simulated GOTO to line 1), and append line $n+1$ containing the instruction Stop. v(WHILE ONE = TWO DO S_1) is the program beginning with a new line number 1, containing the instruction

Jv(ONE), v(TWO), $n+1$, followed by the lines of the revised $v(S_1)$. This program first tests to see if ONE and TWO have the same values. If they do, it jumps to the last line, a halt instruction. If they do not, it proceeds to execute the instructions associated with S_1 (moved down one line), then executes a jump back to line 1 to begin the loop again.

Exercise 5.7.2. Suppose that v has been defined on the statement S_1, and ONE and TWO are identifiers. Give a definition of v(WHILE ONE < TWO DO S_1). *Hint*: In §5.2 we give a SHORTER module.

There are several other while-loop cases since there are other kinds of terms besides identifiers. But by assigning values of terms to identifiers appropriately, they can be reduced to those above. We omit further discussion of them.

Exercise 5.7.3. Show how to define v on IF C THEN S_1 ELSE S_2, provided v has been defined on S_1 and on S_2.

Exercise 5.7.4. Show how to define v on FOR $N := t$ TO u DO S_1, provided v has been defined on S_1.

Thus, finally, $v(S)$ itself has been defined.

Exercise 5.7.5. Use the method above and translate the IMP(L) statement given in Exercise 5.6.1 into a register machine program.

Extending the translation to turn IMP programs into modules is now trivial. Say we have a program with header OUTPUT (IN_ONE, IN_TWO, . . . , IN_LAST): with n inputs, and with program body S. Then $v(S)$ has been defined and, as usual, we assume the only Stop instruction it contains is the final one. The *module* we want has input specification number n and a program part beginning with enough copy instructions to move the contents of $R_1, \ldots, R_n$ into v(IN_ONE), . . . , v(IN_LAST) respectively, followed by the instructions of $v(S)$, appropriately renumbered to allow for the copy instructions that have been added, with its final Stop replaced by a copy instruction to move the contents of v(OUTPUT) into R_1, and with a new final Stop appended.

5.8 ENHANCING AND RESTRICTING IMP

We make a few additions to the IMP machinery to make programming still easier. And we introduce an important sublanguage of IMP.

One IMP addition is a trivial one. We introduce a NULL statement, defined as $X := X$ for some (noninput) identifier X. Obviously, executing NULL changes nothing. Then we allow IF C THEN S as an abbreviation for IF C THEN S ELSE NULL.

The major IMP enhancement is that from now on we allow the use of *nonrecursive functions calls*. For example, if we have written an IMP(L) program with header OUTPUT(INPUT), then in writing a later IMP program P we will allow, say, $A :=$ OUTPUT(B) to be used as another kind of assignment statement. Similarly for multiple-input programs. We take this as an abbreviation for an IMP(L) program. To unabbreviate an occurrence of $A :=$ OUTPUT(B) in program P, rewrite the OUTPUT program body by replacing all occurrences of INPUT by occurrences of B, all occurrences of OUTPUT by occurrences of A, and changing all other identifiers into ones that do not occur in P. Then replace $A :=$ OUTPUT(B) in P by this rewritten OUTPUT body (a single statement). In short, we think of a nonrecursive function call as shorthand for the insertion of the procedure body itself, suitably rewritten to avoid identifier conflict, and to link up inputs and outputs. Since nonrecursive procedure calls can be translated away, they do not add to computational power, but they do make programming more convenient.

There are many other enhancements to IMP(L) that might be introduced without increasing computational abilities. We leave a few to you.

Exercise 5.8.1. We only allowed very simple terms, for example, BUTFIRST($X + Y$) was not allowed. Show how assignment statements involving compound terms could be translated away.

Exercise 5.8.2. Compound conditions are those built up from the simple conditions that we allowed using AND, OR, and NOT. Show how the use of compound conditions in if-then-else statements could be simulated in IMP(L). Then extend the result to while loops.

In the other direction, we want a *less powerful* version of IMP. Using while loops, it is easy to write programs that never halt. But it can be proved that if while loops are not used, programs must terminate. We do not show this here, but it serves as motivation for seeing what can be done without using while loops.

Definition. The programming language IMP$^-$(L) is the sublanguage of IMP(L) without the while loop construction.

Obviously any IMP$^-$(L) programmable function is also IMP(L) programmable. The converse is not true (see Exercises 5.12.4, 5.12.5, and the comments following). We allow the use of if-then-without-else, and nonrecursive functions calls in IMP$^-$(L) just as we do in IMP(L).

Exercise 5.8.3. In Exercise 5.6.5 you were asked to write an IMP(L) program for a FIRST function. Redo that exercise in IMP$^-$(L).

From now on, until further notice, all programs will be in IMP$^-$(L), rather than in IMP(L). Here is an example, a subword function corresponding to

SUBSTR in PL/I, MID$ in BASIC and COPY as often found in Pascal's. The idea is, if X is a word and N and K are numbers, SUBWORD(X, N, K) is the subword of X, beginning at position N, of length K. If a character called for would be located beyond the end of X, or before the start, the output is X unchanged. We make use of the FIRST program you were asked to write in Exercise 5.6.5 and 5.8.3.

```
SUBWORD(X, N, K):
    BEGIN
    N_NUM: = LENGTH(N);         (*make sure we have numbers*)
    K_NUM: = LENGTH(K);
    IF N_NUM = 0̄ THEN           (*test for nonsense input*)
        SUBWORD: = X            (*if nonsense, output X*)
    ELSE
        BEGIN
        N_NUM: = BUTFIRST(N_NUM);
                                (*substring starts after N − 1 letters*)
        MAX: = N_NUM + K_NUM;
        IF X < MAX THEN         (*is X long enough*)
            SUBWORD: = X        (*if X too short, output X*)
        ELSE
            BEGIN
            X_TEMP: = X;
            FOR I: = 1 TO N_NUM DO
                X_TEMP: = BUTFIRST(X_TEMP);
                                (*locate substring start*)
            SUBWORD: = Δ;       (*initialize*)
            FOR J: = 1 TO K_NUM DO
                BEGIN           (*build up substring output*)
                LETTER: = FIRST(X_TEMP);
                SUBWORD: = SUBWORD + LETTER;
                X_TEMP: = BUTFIRST(X_TEMP)
                END             (*J loop*)
            END                 (*second else*)
        END                     (*first else*)
    END.
```

Exercise 5.8.4. Write an IMP$^-$(L) program to reverse words.

EXERCISE 5.8.5. Write IMP$^-$(L) programs for the following functions on number representatives:

(a) multiplication;
(b) exponentiation;
(c) the quotient of a division;

(d) the remainder of a division;
(e) 1 if x divides y, Δ otherwise;
(f) 1 if x is prime, Δ otherwise.

Exercise 5.8.6. Let f and g be two (perhaps partial) functions from L^* to itself, and suppose there are IMP(L) programs to compute each of them. Show that an IMP(L) program exists to compute the composition function $h(x)=f(g(x))$. (Note Exercise 1.5.4.)

5.9 A LOGIC-BASED LANGUAGE LOG(L)

We want to implement the language EFS(**str**(L)) on a register machine. As a tool for this we have introduced the language IMP(L) as an alternate (and easier) way of programming a register machine. It is convenient also to have an alternative to EFS(**str**(L)), based on the logical notation of Chapter 2. We present a programming language we call LOG(L) in which, instead of instructions for *generating* a relation, we give a *description* of it. LOG(L) is essentially a data-base query language. Ultimately we show these very different language styles, EFS(**str**(L)), IMP(L), LOG(L), and register machine language over L, are equivalent in computational power. The equivalence of EFS(**str**(L)) and LOG(L) is established in §5.10; the other equivalences later on.

In what follows let L be an alphabet, with our usual convention that it does not contain any letters that could be confused with any special symbols, now meaning symbols used to express logical relationships, or to write register machine programs, IMP programs, or EFS procedures.

We define a language we call LOG(L). To begin with, we have *identifiers* and *variables* that are exactly the same as in EFS(**str**(L)). *Constants* are words of L^*. *Terms* are variables and constants, again as in EFS(**str**(L)). Finally, *atomic statements* are also defined as they were in EFS(**str**(L)).

The *formulas* of LOG(L) are the expressions built up from atomic statements using $\wedge$, $\vee$, $\exists$ and bounded universal and existential string quantifiers, as in Chapter 2, §2.6 and §2.7. This time, though, we give a more formal definition. At the same time, we define what it means to be a *free occurrence* of a variable. The intention is, the free occurrences are those not covered by quantifiers.

Definition. Of *formula* and *free occurrence*, for LOG(L).

1. An atomic statement is a formula. All occurrences of variables in an atomic statement are free.
2. If X and Y are formulas, so are $(X \wedge Y)$ and $(X \vee Y)$. The free occurrences of variables in these formulas are those occurrences that were free in X together with those occurrences that were free in Y.
3. If X is a formula and v is a variable then $(\exists v)X$ is a formula. Its free occurrences of variables are those that were free in X, except for occurrences of v.

4. If X is a formula, x is a variable, and t is a term, then $(\exists x \text{ SUBWORD } t)X$ and $(\forall x \text{ SUBWORD } t)X$ are formulas. The free variable occurrences in these formulas are the free variable occurrences of X, except for occurrences of x, but together with the added occurrence of t if t is a variable.

Variable occurrences that are not free are said to be *bound*.

Exercise 5.9.1. Verify that the following are formulas of LOG(L), and identify the free variable occurrences. Assume that $A(x, y, z)$ and $B(x, y)$ are atomic statements.

(a) $(\exists x \text{ SUBWORD } y)(A(x, y, z) \wedge B(x, y))$
(b) $(\exists x \text{ SUBWORD } y)A(x, y, z) \wedge B(x, y)$

Definition. If X is a formula, v is a variable, and w is a word of L^*, we use the notation $X(v/w)$ for the formula that results from X when all free occurrences of v are replaced by occurrences of w.

Exercise 5.9.2. Under the same assumptions as in Exercise 5.9.1, what are:

(a) $[(\exists x \text{ SUBWORD } y)(A(x, y, z) \wedge B(x, y))](x/w)$
(b) $[(\exists x \text{ SUBWORD } y)A(x, y, z) \wedge B(x, y)](x/w)$

Next, we give a characterization of *truth* for formulas of LOG(L) *that contain no free variable occurrences*. We used this notion in Chapter 2, §2.6 and §2.7. Our formal characterization now is only superficially different. We use the notion of *work space* exactly as with EFS languages. The *basic* work space for LOG(L) is the same as for EFS(**str**(L)), with CON reserved, and representing $\mathbf{CON}_L$. Any other work space to be considered is an expansion of this one.

Definition. Of *truth* for formulas of LOG(L) without free variable occurrences and with all identifiers reserved, in work space **W**.

1. Let RES be a reserved identifier of work space **W**, representing an n-place relation, and let $w_1, \ldots, w_n$ be words of L^*. The atomic formula RES$(w_1, \ldots, w_n)$ is true in **W** if $\langle w_1, \ldots, w_n \rangle$ is in the relation that RES represents in **W**.
2. Let X and Y be formulas without free variable occurrences. $(X \wedge Y)$ is true in **W** if both X and Y are true in **W**. $(X \vee Y)$ is true in **W** if at least one of X or Y is true in **W**.
3. Let $(\exists v)X$ be a formula without free variable occurrences. $(\exists v)X$ is true in **W** provided, for some word w of L^*, $X(v/w)$ is true in **W**.
4. Let $(\exists x \text{ SUBWORD } c)X$ and $(\forall x \text{ SUBWORD } c)X$ be formulas without free variable occurrences (hence c is a word of L^*). $(\exists x \text{ SUBWORD } c)X$ is true in **W** provided, for some word w that is a subword of c, $X(x/w)$ is

true in **W**. ($\forall x$ SUBWORD $c)X$ is true in **W** provided, for every word w that is a subword of c, $X(x/w)$ is true in **W**.

Exercise 5.9.3. Use the basic work space of LOG($\{a, b\}$). Which of the following are true?

(a) $(\exists x)(\exists y)(\text{CON}(y, ab, x) \wedge \text{CON}(x, b, babb))$
(b) $(\exists x)(\exists y)(\text{CON}(ab, x, y) \wedge \text{CON}(ab, y, x))$.

The following notational convention will be used from now on; it should be familiar from Chapter 2. If F is a formula, we write $F(x_1, \ldots, x_n)$ to indicate that the free variables of F are among $x_1, \ldots, x_n$, and for words $w_1, \ldots, w_n$ we write $F(x_1/w_1)\ldots(x_n/w_n)$ as $F(w_1, \ldots, w_n)$. Please note that if we write $F(x_1, \ldots, x_n)$, it does not require that each of $x_1, \ldots, x_n$ actually *has* free occurrences in F. If x does not occur free in F, $F(x/w) = F$.

If $F(x_1, \ldots, x_n)$ is a formula of LOG(L) all of whose identifiers are reserved in **W**, we can think of it as specifying a relation relative to the work space **W**, the relation whose members are the n-tuples for which F is true. "Programming" in LOG(L) consists of giving such a relation a name, and enlarging the work space with it.

Definition. Let **W** be a work space and let $F(x_1, \ldots, x_n)$ be a formula of LOG(L) all of whose atomic formulas use reserved identifiers of **W**. Further, let NAME be an identifier not reserved in work space **W**. The expression

$$\text{NAME}(x_1, \ldots, x_n) \Leftrightarrow F(x_1, \ldots, x_n)$$

is a *definition* of NAME in work space **W**. We expand **W** by adding NAME as a reserved identifier, specifying that it represents the relation $\{\langle w_1, \ldots, w_n\rangle \mid w_1, \ldots, w_n \in L^* \text{ and } F(w_1, \ldots, w_n) \text{ is true in } \mathbf{W}\}$.

For example, start with the basic work space of LOG(L). Use the following definition.

$$\text{BEGINS}(x, y) \Leftrightarrow (\exists z)\text{CON}(x, z, y)$$

We now expand the work space by adding BEGINS as a reserved identifier, where BEGINS represents

$$\{\langle w_1, w_2\rangle \mid w_1, w_2 \in L^* \quad \text{and} \quad (\exists z)\text{CON}(w_1, z, w_2)\}.$$

The members of this relation are those pairs $\langle w_1, w_2\rangle$ where word w_1 is the beginning part of w_2. Another simple example, with obvious intuitive interpretation is

$$\text{EQUAL}(x, y) \Leftrightarrow \text{CON}(x, \Delta, y).$$

There is a significant difference between $(\exists x)X$ and ($\exists x$ SUBWORD $y)X$. The bounded quantifier "asks" if a certain x exists, and specifies where we should look: through the subwords of y. This is a search with an end to it, since any given word has only a finite number of subwords. $(\exists x)X$, on the other hand, merely asks if an x meeting condition X exists, and places no other restrictions. The search for such an x is an *unbounded search.* The possibility of such things in LOG(L) is the counterpart, in conventional programming languages, of loops that may never exit, such as while loops. So it is of importance to us to see how much "programming" can be done without using the unbounded existential quantifier. Consequently we introduce a weaker sublanguage containing only bounded quantifiers, and give definitions in it, moving to LOG(L) only when forced to.

Definition. The language LOG$^-$(L) is like LOG(L) except that unbounded existential quantifiers are not allowed.

For example, BEGINS above can be redone to fit into the LOG$^-$(L) framework, as follows.

$$\text{BEGINS}(x, y) \Leftrightarrow (\exists z \text{ SUBWORD } y)\text{CON}(x, z, y).$$

Since the definition of EQUAL involved no quantifiers, it is also a LOG$^-$(L) definition. Finite relations have trivial definitions. For instance, suppose the alphabet L is $\{a, b, c\}$. Then we might write the definition

$$\text{LETTER}(x) \Leftrightarrow \text{EQUAL}(x, a) \vee \text{EQUAL}(x, b) \vee \text{EQUAL}(x, c).$$

The preceding is not really a definition because the expression on the right is not a formula. We should have written something like ((EQUAL(x, a) $\vee$ EQUAL(x, b)) $\vee$ EQUAL(x, c)). However, we will leave parentheses out, to make reading easier, when it is clear how they could be restored. The technique used for LETTER applies to any finite relation. So we can assume we have a definition for DIFFERENT_LETTER, with the obvious intuitive meaning.

The next examples are a little more complicated, and begin to show the inherent richness of LOG$^-$(L).

$$\text{PROPERLY\_BEGINS}(x, y) \Leftrightarrow (\exists a \text{ SUBWORD } y)(\exists w \text{ SUBWORD } y)$$
$$(\text{LETTER}(a) \wedge \text{CON}(x, a, w) \wedge \text{BEGINS}(w, y))$$

$$\text{NOT\_EQUAL}(x, y) \Leftrightarrow \text{PROPERLY\_BEGINS}(x, y) \vee$$
$$\text{PROPERLY\_BEGINS}(y, x) \vee (\exists a \text{ SUBWORD } x)(\exists b \text{ SUBWORD } y)$$
$$(\exists w \text{ SUBWORD } x)(\exists u \text{ SUBWORD } x)(\exists v \text{ SUBWORD } y)$$
$$(\text{DIFFERENT\_LETTER}(a, b) \wedge \text{CON}(w, a, u) \wedge \text{CON}(w, b, v) \wedge$$
$$\text{BEGINS}(u, x) \wedge \text{BEGINS}(v, y))$$

To understand the biggest clause of NOT_EQUAL, keep in mind the following: $u = wa$ and $x = u\ldots = wa\ldots$, $v = wb$ and $y = v\ldots = wb\ldots$.

Exercise 5.9.4. Write definitions for the following in $LOG^{-}(L)$.

(a) ENDS(x, y) word x ends word y.
(b) PROPERLY_ENDS(x, y) word x properly ends word y.
(c) NOT_SUBWORD(x, y) word x is not a subword of word y.
(d) SUBWORD_BUT_NOT_BEGINS(x, y) word x is a subword of word y but does not begin it.
(e) NOT_BEGINS(x, y) word x does not begin word y.
(f) NOT_ENDS(x, y) word x does not end word y.
(g) INSIDE(x, y) word x is a subword of word y but does not begin or end it.
(h) NOT_INSIDE(x, y) word x is not a subword of word y, or is but begins or ends it.

5.10 EFS AND LOG

We prove the equivalence of EFS- and LOG-style languages for specifying relations on words. We do not treat EFS procedures that accept inputs, largely in the interests of a shorter book. The techniques presented here extend readily, however.

For the rest of this section L is a fixed alphabet. One direction of the translation is immediately available.

Proposition 5.10.1. *Let **R** be an* n*-place relation on* L**. If there is a LOG*(L) *definition for **R**, then **R** is the output of some EFS*(**str**(L)) *procedure.*

Proof. Every given relation of the basic work space is generated in EFS(**str**(L)), trivially, since only **CON** is involved. Now suppose that we have a work space **W** all of whose given relations are generated in EFS(**str**(L)). We showed in Chapter 2 that the various ways of forming more complex formulas from simpler ones all corresponded to operations that turn generated relations into generated relations. Consequently, any LOG definable relation **R** in **W** is generated by some EFS procedure. Then all given relations in any LOG expansion of **W** are generated. And so on.

The other direction is more work. We use EFS procedure traces, but there is a difficulty involved. A typical line of a trace will be a substitution instance of a procedure statement. This itself is a word, of course, but one that involves letters beyond the alphabet L. Specifically it may involve arrows and commas and the letters used to form identifiers. This suggests we need to consider a larger alphabet if we are to treat traces themselves as computational objects. But we made the convention that we would never use as letters of an alphabet symbols playing a special role. The way out is simple; we add *new* letters to the alphabet L. The arrow, for instance, is not one of these new letters, but there will be a letter that we can treat as if it were the arrow. And similarly for all special symbols. Specifically, we introduce the following new letters.

⇛	used as a "stand in" for	$\rightarrow$
₀	used as a "stand in" for	,
⟦, ⟧	used as "stands ins" for	(,)
𝔸, . . . , ℤ, ₌	used as "stands ins" for	$A, \ldots, Z,$_
#	used as a "stand in" for	new line

In displaying EFS traces, we presented different statements on different lines. The symbol # is to symbolize the starting of a new line.

We denote by E the alphabet of extra symbols above. By design, E contains no symbols that have a special role in writing EFS procedures or LOG definitions. We further assume that E contains no symbol already in L. Let L' be the alphabet $L \cup E$. Then any EFS statement without variables *over* L has its counterpart *in* L'. For instance, say $L = \{a, b\}$, and consider the EFS statement

CON(a, b, ab)$\rightarrow$OUT(ab).

Corresponding to this is the word *of* L'^*

ℂ𝕆ℕ⟦a_0b_0ab⟧⇛𝕆𝕌𝕋⟦ab⟧.

We set up a series of LOG definitions for words of L'^*, definitions that say the counterparts of EFS statements behave like them. Our immediate goal is a LOG definition of the counterpart of an EFS trace.

We begin with some general LOG definitions. To start, we have the basic work space of LOG(L') = LOG($L \cup E$), extended using the various definitions of the previous section, in particular those you wrote in Exercise 5.9.4. We further extend that step by step. We give definitions in the restricted language LOG$^-$(L') as far as possible, not using the unbounded existential quantifier. As a matter of fact, all but one definition in this section will meet this restriction.

LETTER_OF_L. This is a finite set, hence a definition can be given by just enumerating the letters.

L is a subalphabet of L'. We want a definition of the set of words of L^* alone, because they are used to get substitution instances of procedure statements. The idea is, w is a word over L if every nonempty subword of w contains a letter of L, that is, has a letter of L as a subword.

WORD_OVER_L(x)$\Leftrightarrow$($\forall y$ SUBWORD x)(EQUAL(y, Δ) $\vee$
($\exists z$ SUBWORD y)LETTER_OF_L(z))

Our final preliminary items concern concatenation. We have CON available as a 3-place relation. But things will get hard for us to read if this is all we have. We define a sequence of more encompassing concatenation relations, still using CON as identifier. Understand, the sequence of definitions is not infinite. Rather, it includes as many items as we need for a particular application under consideration. This is always a finite number of definitions, but how many depends on circumstances. In each defined relation, the last item in the tuple is the concatenation of all those listed earlier. Technically, the re-use of the identifier CON is not legal, but it is convenient and will cause no harm. To be proper we could use CONONE, CONTWO, etc. instead.

$\mathrm{CON}(x, y, z, v) \Leftrightarrow (\exists u \text{ SUBWORD } v)(\mathrm{CON}(x, y, u) \wedge \mathrm{CON}(u, z, v))$
$\mathrm{CON}(w, x, y, z, v) \Leftrightarrow (\exists u \text{ SUBWORD } v)(\mathrm{CON}(w, x, y, u) \wedge \mathrm{CON}(u, z, v))$
etc.

Next, suppose that A_i is some particular EFS procedure statement. There is a corresponding $\mathrm{LOG}^-(L')$ definition for its substitution instances over L, with $\Rightarrow$ used for $\rightarrow$, $_0$ for comma, and so on. To demonstrate this, it is simplest to present an illustrative example. Say that A_i is the EFS(**str**(L)) procedure statement

$\mathrm{ONE}(x) \rightarrow \mathrm{TWO}(x, ab, y) \rightarrow \mathrm{THREE}(y)$

where x and y are variables and a and b are letters of L. In this case, we use the following $\mathrm{LOG}^-(L')$ definition.

INSTANCE_OF_A_i_OVER_$L(w) \Leftrightarrow$
  $(\exists x \text{ SUBWORD } w)(\exists y \text{ SUBWORD } w)(\mathrm{WORD\_OVER\_}L(x) \wedge$
    $\mathrm{WORD\_OVER\_}L(y) \wedge$
    $\mathrm{CON}(\mathbb{ONE}[\![, x,]\!] \Rightarrow \mathbb{TWO}[\![, x, {}_0ab_0, y,]\!] \Rightarrow \mathbb{THREE}[\![, y,]\!], w))$

This says that w is the result of concatenating the various special symbols of A_i, but with words over L used in place of variables. From now on, we assume that, when given a procedure statement of EFS(**str**(L)), we can write out a corresponding $\mathrm{LOG}^-(L')$ definition for its substitution instances, as we did above.

Next, we can assume that when working with EFS(**str**(L)), we are using the basic work space. (See the discussion in Chapter 1, §1.5.) Then only the (3-place) concatenation relation is given. We need a $\mathrm{LOG}^-(L')$ definition for the instances of it that can appear in traces.

CON_INSTANCE_OVER_$L(w) \Leftrightarrow$
  $(\exists x \text{ SUBWORD } w)(\exists y \text{ SUBWORD } w)(\exists z \text{ SUBWORD } w)$
    $(\mathrm{WORD\_OVER\_}L(x) \wedge \mathrm{WORD\_OVER\_}L(y) \wedge$
    $\mathrm{WORD\_OVER\_}L(z) \wedge$
    $\mathrm{CON}(\mathbb{CON}[\![, x, {}_0, y, {}_0, z,]\!], w) \wedge \mathrm{CON}(x, y, z))$

Different lines of a trace are separated by the creation of a new line; when working over L', $\#$ stands for this. In order not to make special cases of beginnings and ends of traces, we will assume that $\#$ appears at the start and at the end as well. Then, a line (of a trace) can be recognized as a string that appears between occurrences of $\#$ and does not, itself, contain an occurrence of $\#$. The following is intended to characteize the relation, x is a line of y, in this sense.

$\mathrm{LINE\_OF}(x, y) \Leftrightarrow (\exists w \text{ SUBWORD } y)(\mathrm{NOT\_INSIDE}(\#, w) \wedge$
  $\mathrm{CON}(\#, x, \#, w))$

also

NOT_LINE_OF(x, y) ⇔ (∀w SUBWORD y)(INSIDE(#, w) ∨
(∃z SUBWORD w)(CON(#, z, #, w) ∧ NOT_EQUAL(z, x)).

The next item is a definition of being an *earlier* line. This comes up in stating the Assignment Rule for EFS. The following is intended to characterize the relation, x is an earlier line than y in z.

EARLIER_LINE(x, y, z) ⇔
LINE_OF(x, z) ∧ LINE_OF(y, z) ∧
(∃w SUBWORD z)(PROPERLY_BEGINS(x, w)
∧ PROPERLY_ENDS(y, w)).

Now we are ready to deal with traces themselves. Or, rather, it will be simpler for us to use the restricted traces of Chapter 1, §1.4. Recall that these use the Restricted Assignment Rule in which we only consider one condition of a conditional assignment at a time. To use it, or for that matter, to use the full Assignment Rule, we need to be able to identify atomic procedure statements. Well, a line of a trace is atomic if it does not contain the arrow symbol.

ATOMIC(x) ⇔ NOT_SUBWORD(⇒, x).

Proposition 5.10.2. *Let the following be an EFS(**str**(L)) procedure in the basic work space:*

IDENTIFIER (n):
A_1;
A_2;
⋮
A_k.

Let T *be the set of restricted traces for this procedure, written with ⇒ for →, o for comma, and so on. There is a LOG⁻(L′) definition of* T.

Proof. The definition, which exactly mirrors that of restricted trace from Chapter 1, §1.4, is as follows.

TRACE(w) ⇔ BEGINS(#, w) ∧ ENDS(#, w) ∧
(∀x SUBWORD w)(NOT_LINE_OF(x, w) ∨
(LINE_OF(x, w) ∧
(CON_INSTANCE_OVER_L(x) ∨
INSTANCE_OF_A_1_OVER_L(x) ∨
. . . ∨
INSTANCE_OF_A_k_OVER_L(x) ∨
(∃u SUBWORD w)(∃v SUBWORD w)
(EARLIER_LINE(u, x, w) ∧
EARLIER_LINE(v, x, w) ∧
ATOMIC(u) ∧
CON(u, ⇒, x, v)))).

Thus far, all definitions have been in $\mathrm{LOG}^-(L')$. Now, finally, we need an unbounded existential quantifier.

Corollary 5.10.3. *If* **R** *is generated in EFS*(**str**(L)) *then* **R** *has a definition in LOG*(L′).

Proof. Suppose **R** is generated in EFS(**str**(L)). Say that **R** is the output of the procedure

IDENTIFIER (n):
    (procedure body of IDENTIFIER).

in the basic work space. We get the members of **R** by looking at lines of restricted traces of the form IDENTIFIER($w_1, \ldots, w_n$). Then the following LOG(L') definition will serve, in which we use the definition TRACE from Proposition 5.10.2.

$R(x_1, \ldots, x_n) \Leftrightarrow (\exists w)(\mathrm{TRACE}(w) \wedge$
    $(\exists z$ SUBWORD $w)(\mathrm{CON}(\mathbb{IDENTIFIER}[\![, x_1, {}_0, \ldots, {}_0, x_n,]\!], z) \wedge$
    LINE_OF(z, w)).

It is true that, in moving from EFS to LOG we were forced to increase the size of the alphabet. This can be avoided. When working with EFS string data structures, we showed how to code a larger alphabet into a smaller one (Chapter 4, §4.5, 4.6). A similar device works for LOG definability. Presenting the details would get us too involved in something we have no particular need for, so we merely remark that LOG(L) can be substituted for LOG(L') in Corollary 5.10.3.

The Corollary proof actually establishes a stronger result than was stated. Not only is every generated relation of EFS(**str**(L)) definable in LOG(L'), but *only a single unbounded existential quantifier is needed.* This is related to Theorem 3.13.1, which essentially said that only a single use of recursion was needed in EFS procedures. It is further related to a result about while loops which we will give in §5.12.

Exercise 5.10.1. Use the method of this section and produce a LOG definition corresponding to the following simple EFS(**str**(L)) procedure, where $L = \{1\}$.

EVEN (1):
    EVEN(Δ);
    EVEN(x)$\rightarrow$CON(x, 11, y)$\rightarrow$EVEN(y).

5.11 LOG⁻ AND IMP⁻

EFS and LOG languages manipulate relations while IMP uses functions. We can treat a function as a relation, by using its graph. Now we need a device for treating a relation as a function.

Definition. Let $\mathbf{R}$ be an n-place relation. *The characteristic function* of $\mathbf{R}$ is the n-ary function $c_{\mathbf{R}}$ given by

$$c_{\mathbf{R}}(x_1, \ldots, x_n) = \begin{cases} \bar{1} & \text{if} \quad \langle x_1, \ldots, x_n \rangle \in \mathbf{R} \\ \bar{0} & \text{if} \quad \langle x_1, \ldots, x_n \rangle \notin \mathbf{R} \end{cases}$$

A characteristic function tells us when a relation holds and when it doesn't, which is the essential information about a relation. We have used number representatives $\bar{1}$ and $\bar{0}$, because we want to keep things in the realm of words. Think of $\bar{1}$ as *true* and $\bar{0}$ as *false*.

Definition. If an IMP(L) program exists that computes a function f, we call f *IMP(L) computable.* If the characteristic function of a relation $\mathbf{R}$ is IMP(L) computable, we say the *relation* $\mathbf{R}$ is IMP(L) computable. We use similar definitions for IMP$^-$(L) computable.

As we saw in §5.7, the IMP(L) computable functions are among those computable using register machines. Since register machine computation translates into EFS computation (Corollary 5.5.2), once we show that EFS(**str**(L)) can be interpreted in IMP(L), the equivalence of IMP(L) computability and register machine computability follows. The main result of this section is really a lemma for the interpretation of EFS into IMP. We state it now, the remainder of the section contains its proof, and the next section contains its major application.

Proposition 5.11.1. Let $\mathbf{R}$ be a relation on L^*. If $\mathbf{R}$ has a LOG$^-$(L) definition then $\mathbf{R}$ is IMP$^-$(L) computable.

The proof is by induction on the complexity of the LOG$^-$(L) definition of $\mathbf{R}$. What we show in the remainder of the section is this. The atomic formulas of LOG$^-$(L) define IMP$^-$(L) computable relations. And each of the methods for building more complex formulas of LOG$^-$(L), by using $\wedge$, $\vee$, and bounded quantification, turn IMP$^-$(L) computable relations into IMP$^-$(L) computable relations.

We begin with atomic formulas, and we assume we are in the basic work space. Then CON is the identifier, but there are still several cases to be considered. For example, suppose x, y, and z are distinct variables and c and d are constants, words over L. Then the following are atomic formulas, but the behavior of each is different than that of the rest: CON(x, y, z), CON(x, x, y), CON(x, c, z), CON(x, c, d), etc. We will write an IMP$^-$(L) program for the characteristic function of the first of these. The rest are more or less similar.

An IMP$^-$(L) program for the characteristic function of the relation defined by CON(x, y, z).

```
ATOMIC_ONE(X, Y, Z):
    BEGIN
    W: = X + Y;
    IF W = Z THEN ATOMIC_ONE: = 1̄
    ELSE ATOMIC_ONE = 0̄
    END.
```

Example 5.11.1. Write $\text{IMP}^-(L)$ programs for the relations defined by $\text{CON}(x, x, y)$ and $\text{CON}(x, c, z)$, each of which is a 2-place relation.

Next, we show that each way of forming a more complicated $\text{LOG}^-(L)$ formula from simpler ones corresponds to an $\text{IMP}^-(L)$ construction. There are four cases to be considered: $\wedge$, $\vee$, and the two bounded quantifiers.

Let R and S be $\text{LOG}^-(L)$ formulas. They may not have all their free variables in common. We symbolize this by writing $\text{R}(x_1, \ldots, x_i, z_1, \ldots, z_k)$ and $\text{S}(y_1, \ldots, y_j, z_1, \ldots, z_k)$. The z's are the common free variables, while the x's only occur in R and the y's only in S. Suppose we already have $\text{IMP}^-(L)$ programs that compute the characteristic functions of the relations that R and S define. Say these programs are

```
R(X1, . . . , Xi, Z1, . . . , Zk):
    (body of R)
```

and

```
S(Y1, . . . , Yj, Z1, . . . , Zk):
    (body of S)
```

Here is an $\text{IMP}^-(L)$ program that computes the characteristic function of $\text{R} \wedge \text{S}$.

```
R_AND_S(X1, . . . , Xi, Y1, . . . , Yj, Z1, . . . , Zk):
    BEGIN
    TEST_R: = R(X1, . . . , Xi, Z1, . . . , Zk);
    TEST_S: = S(Y1, . . . , Yj, Z1, . . . , Zk);
    IF TEST_R = 0̄ THEN R_AND_S: = 0̄
    ELSE
        IF TEST_S = 0̄ THEN R_AND_S: = 0̄
        ELSE R_AND_S: = 1̄
    END.
```

Example 5.11.2. Under the same assumptions as above, write an $\text{IMP}^-(L)$ program to compute the characteristic function for $\text{R} \vee \text{S}$.

Next we turn to bounded quantifiers. We make use of the program DIFF, for subtraction, given in §5.6. Suppose that $\text{R}(x_1, \ldots, x_n, y)$ is a $\text{LOG}^-(L)$ formula and we already have an $\text{IMP}^-(L)$ program to compute the characteristic function of the relation that R defines. Say the program is

```
R(X1, . . . , Xn, Y):
     (program body)
```

Consider the formula $(\forall y \text{ SUBWORD } z)\text{R}(x_1, \ldots, x_n, y)$. We give an $\text{IMP}^-(L)$ program for its characteristic function. We use SUBWORD from §5.8 to extract subwords, and loops to run through all of them. If z is the empty word, $(\forall y \text{ SUBWORD } z)\text{R}(x_1, \ldots, x_n, y)$ is equivalent to $\text{R}(x_1, \ldots, x_n, \Delta)$. The program checks for this first.

```
UNIV(X1, . . . , Xn, Z):
     IF Z = Δ THEN UNIV: = R(X1, . . . , Xn, Δ)
     ELSE
          BEGIN
          UNIV: = 1̄;       (*assume true, change if found false*)
          L: = LENGTH(Z);
          FOR K: = 1̄ TO L DO
               BEGIN      (*try all subwords starting at position K*)
               E: = DIFF(L,K);
               E: = E + 1̄; (*longest subword of Z starting at K has length E*)
               FOR J: = 0̄ TO E DO
                    BEGIN (*try subword of length J*)
                    W: = SUBWORD(Z,K,J);
                    TEST: = R(X1, . . . , Xn, W);
                    IF TEST = 0̄ THEN UNIV: = 0̄
                              (*counterexample found*)
                    END       (*J loop*)
               END            (*K loop*)
          END                 (*else*)
```

Exercise 5.11.3. Write a similar $\text{IMP}^-(L)$ program to take care of the bounded existential quantifier.

We have dealt with all the cases and the proof of Proposition 5.11.1 is finished. Once again we have given a constructive proof, a recipe for writing an $\text{IMP}^-(L)$ program for a characteristic function when given a $\text{LOG}^-(L)$ definition.

Exercise 5.11.4. Convert the $\text{LOG}^-(L)$ definition of BEGINS in §5.9 into an $\text{IMP}^-(L)$ program for its characteristic function.

5.12 PROGRAMMING LANGUAGE EQUIVALENCES

If something is an output of an EFS procedure we can be convinced by seeing an appropriate trace. To be convinced that something is not an output of an EFS procedure, some other technique must be used; some suitable model could be created, for instance. Being in, and not being in, a generated relation are not symmetric situations. We have been treating relations via functions by using characteristic functions, defined in §5.11. Such functions give us full information about what is in a given relation and what what is not. Now we need something that is less symmetric in its behavior.

Definition. Let **S** be an n-place relation. The *semicharacteristic function* of **S** is the n-ary partial function $s_{\mathbf{S}}$ given by

$$s_{\mathbf{S}}(x_1, \ldots, x_n) = \begin{cases} \bar{1} & \text{if } \langle x_1, \ldots, x_n \rangle \in \mathbf{S} \\ \text{undefined} & \text{if } \langle x_1, \ldots, x_n \rangle \notin \mathbf{S} \end{cases}$$

If $\langle x_1, \ldots, x_n \rangle$ is in the relation **S**, the semicharacteristic function $s_{\mathbf{S}}$ will "tell" us that. If $\langle x_1, \ldots, x_n \rangle$ is not in **S**, the semicharacteristic function gives us no information at all. If you think of $s_{\mathbf{S}}$ as being computed by machine, not getting an output after a certain amount of time does not tell us whether the function being computed is undefined for the input we used, or whether it is defined but the calculation of the output takes longer than we have waited so far. Thus, no output so far really is no information. We will see later on that one cannot expect better in general.

Definition. If an IMP(L) program exists that computes the semicharacteristic function $s_{\mathbf{S}}$ of a relation **S**, we say the relation **S** is *IMP*(L) *semicomputable.*

We could define a notion of IMP$^-$(L) semicomputable, but it is of no real interest, since all IMP(L) programs must halt, thus producing an output. For the rest of this section L is some nonempty alphabet, and L' is the alphabet $L \cup E$, where E consists of the extra letters introduced in §5.10: $\Rightarrow$, $_0$, $[\![$, $]\!]$, $\mathbb{A}, \ldots, \mathbb{Z}$, $_=$, and $\#$. The central result of the section is this.

Proposition 5.12.1. *Let* **S** *be a relation on* L. *If* **S** *is the output of an EFS*(**str**(L)) *procedure, then* **S** *is IMP*(L$'$) *semicomputable.*

Proof. Say **S** is the output of the EFS(**str**(L)) procedure

S_PROCEDURE (n):
(procedure body of S_PROCEDURE).

By Proposition 5.10.2, if T is the set of restricted traces for this procedure, written in the alphabet L', then there is a LOG$^-$(L') definition of T. Say TRACE(w) is such a definition.

$\langle t_1, \ldots, t_n \rangle$ is an output of S_PROCEDURE just when some EFS trace has S_PROCEDURE($t_1, \ldots, t_n$) as a line. This happens just when some word w over L' satisfies the following LOG$^-$(L') definition.

TRACE_SHOWING_OUTPUT($w, t_1, \ldots, t_n$)$\Leftrightarrow$TRACE(w) $\wedge$
  ($\exists z$ SUBWORD w)(CON($\mathbb{S}_{=}\mathbb{PROCEDURE}[\![, t_1, {}_0, \ldots, {}_0, t_n,]\!], z$) $\wedge$
  LINE_OF(z, w))

Then $\langle t_1, \ldots, t_n \rangle$ is an output of procedure S_PROCEDURE if and only if there is some word w over L' such that we have TRACE_SHOWING_OUTPUT($w, t_1, \ldots, t_n$). Since we have a LOG$^-$(L') definition of TRACE_SHOWING_OUTPUT to work with, we can write an IMP$^-$(L') program for its characteristic function, by Proposition 5.11.1. Say this program is

```
TRACE_SHOWING_OUTPUT(W, T1, . . . , Tn):
    (program body)
```

So to "know" that $\langle t_1, \ldots, t_n \rangle$ is an output of S_PROCEDURE, we must search for a value of W for which TRACE_SHOWING_OUTPUT(W, $t_1, \ldots, t_n$) returns the value true. What we need to complete the proof is some IMP(L') program that can, one after the other, generate all possible words over L' and test them using the program above, to see if any will serve as a suitable value for W. There are several ways of generating all words over an alphabet. We use the following device. We get all words of length $k+1$ by adjoining each letter of the alphabet to the end of each word of length k. This requires us to remember all the words of length k when generating words of length $k+1$. This is something we could use a stack for, if we had such a thing available. We do not, so instead we concatenate all the words of length k into a single big word. We are able to retrieve the original words from it by dividing it up into consecutive subwords of length k. In the following program, $a, b, \ldots, \mathbb{A}, \mathbb{B}, \ldots$ is the alphabet L'. Our testing for a trace starts with words of length 2, rather than 1. This is simply a matter of convenience. Now, here is the IMP(L') program that will complete the proof of Proposition 5.12.1. In it we use several procedures writen earlier, including one for multiplication which was part of Exercise 5.8.5.

```
S_PROCEDURE_SEMI_CHAR(T1, . . . , Tn):
    BEGIN
    ALPHABET: = ab. . .𝔸𝔹. . . ;      (*alphabet L'*)
    SIZE: = LENGTH(ALPHABET);
    WORD_LENGTH: = 1̄      (* word length just considered*)
    LIST: = ALPHABET;      (*LIST is to store all words of length
                             WORD_LENGTH concatenated. It is
                             initialized to the length 1 words*)
    WORD_NUMBER: = SIZE;      (*number of words in LIST*)
    S_PROCEDURE_SEMI_CHAR: = 0̄;      (*initialize to false*)
    WHILE(S_PROCEDURE_SEMI_CHAR = 0̄)DO
        BEGIN
        TEMP_LIST: = Δ;      (*TEMP_LIST is all words of length
                               WORD_LENGTH + 1̄ tried so far*)
        BOUND: = BUTFIRST(WORD_NUMBER);
```

```
        FOR I: = 0̄ TO BOUND DO        (*run through words in LIST*)
            BEGIN
            POSITION: = TIMES(I, WORD_LENGTH);
            POSITION: = POSITION + 1̄;       (*locate start of word I*)
            WORD: =
                SUBWORD(LIST, POSITION, WORD_LENGTH);
                (*extract word I from LIST*)
            FOR J: = 1̄ TO SIZE DO       (*SIZE is of L'*)
                BEGIN
                LETTER: = SUBWORD(ALPHABET,J,1̄);
                (*choose a letter of L'*)
                NEW_WORD: = WORD + LETTER;       (*add letter on*)
                TEST: = TRACE_SHOWING_OUTPUT(NEW_WORD,
                                                        T1, . . . , Tn);
                                                (*try NEW_WORD out*)
                IF TEST = 1̄ THEN S_PROCEDURE_SEMI_CHAR: = 1̄;
                                  (*remember suitable trace found*)
                TEMP_LIST: = TEMP_LIST + NEW_WORD;
                          (*remember word just tried*)
                END       (*K loop*)
            END;          (*I loop*)
        WORD_LENGTH: = WORD_LENGTH + 1̄;
                          (*start over with longer words*)
        WORD_NUMBER: = TIMES(WORD_NUMBER, SIZE);
        LIST: = TEMP_LIST
        END               (*WHILE loop*)
    END.
```

The behavior of this IMP(L') program is quite simple. If a word w such that TRACE_SHOWING_OUTPUT($w, t_1, \ldots, t_n$) exists, the program will eventually construct the word, discover it works, exit the while loop and output $\bar{1}$. If no such word exists, the program will continue looping, trying longer and longer words, never halting. Thus S_PROCEDURE_SEMI_CHAR computes the semicharacteristic function of **S**, and hence **S** is IMP(L') semicomputable. The proof is done.

Once again a stronger result was proved than was stated. In the program written above, all constructs are IMP$^-$(L) constructs except for the single while loop. Thus, what was actually shown was:

*If **S** is generated then there is an IMP*(L′) *program that computes the semicharacteristic function of **S** and that involves a single while loop and otherwise only involves the counting loops and other constructs of IMP*$^-$(L).

In §5.10 we saw that only a single unbounded existential quantifier was needed in LOG definitions. And in Chapter 3, §3.13 we saw that only a single application of recursion was needed for EFS definitions. All these are versions of the same result, which goes back to Kleene. He characterized the computable functions quite differently than we did (though equivalently). In his formulation, there was an *unbounded search* operator. The Kleene Normal Form Theorem says, in part, that any computable function can be characterized with a single use of unbounded search, together with the rest of his machinery.

Theorem 5.12.2. *Let* **S** *be a relation on* L*. *The following are equivalent*:

1. ***S*** *is the output of an EFS*(***str***(L)) *procedure (i.e.,* ***S*** *is generated).*
2. ***S*** *has a LOG*(L′) *definition.*
3. ***S*** *is IMP*(L′) *semicomputable.*
4. *The semicharacteristic function of* ***S*** *is register machine computable over* L′.

Proof. (1) and (2) are equivalent by Proposition 5.10.1 and Corollary 5.10.3. (1) implies (3) by Proposition 5.12.1, and (3) implies (4) via the implementation procedure given in §5.7. Finally we show that (4) implies (1). If the semicharacteristic function $s_{\mathbf{S}}$ is register machine computable then by Corollary 5.5.2, the graph of $s_{\mathbf{S}}$ is generated by some EFS(**str**(L')) procedure. Say by

S_GRAPH ($n+1$):
  (procedure body)

where $\langle t_1, \ldots, t_n\rangle \in \mathbf{S}$ if and only if
  $s_{\mathbf{S}}(t_1, \ldots, t_n) = \bar{1}$ if and only if
  $\langle t_1, \ldots, t_n, \bar{1}\rangle$ is an output of S_GRAPH.

The following is obviously an EFS(**str**(L')) procedure for S itself.

$S(n)$:
  S_GRAPH$(X_1, \ldots, X_n, \bar{1}) \rightarrow$ S$(X_1, \ldots, X_n)$.

Then **S** is generated in ESF(**str**(L')). Finally, the work of Chapter 4, §4.5, 4.6 shows that **S** is generated in EFS(**str**(L)).

Definition. Call a relation **R** *semicomputable* if it meets any of the conditions of Theorem 5.12.2.

Theorem 5.12.3. *Let* F *be an* n-*place (partial) function on* L*. *The following are equivalent*:

1. *The graph of* F *is generated in EFS*(***str***(L)).
2. *The graph of* F *has a LOG*(L′) *definition.*
3. F *is IMP*(L′) *computable.*
4. F *is register machine computable over* L′.

Proof. The argument is essentially the same as that for Theorem 5.12.2, except for the proof that (1) implies (3), for which we proceed as follows. Suppose the graph of F is generated in EFS(**str**(L)). By Proposition 5.12.1, that graph is IMP(L') semicomputable, say by the IMP program

GRAPH($X_1, \ldots, X_n$, Y):
(body of program)

That is,

$F(x_1, \ldots, x_n) = y$ if and only if
$\langle x_1, \ldots, x_n, y \rangle$ is in the graph of F if and only if
$\text{GRAPH}(x_1, \ldots, x_n, y) = \bar{1}$.

Then an IMP(L') program for F itself can be constructed along the same lines as the program written in the proof of Proposition 5.12.1. That is, we input to the program the values $x_1, \ldots, x_n$, and have it generate, one after the other, all possible words y and, for each one, test whether $\text{GRAPH}(x_1, \ldots, x_n, y) = \bar{1}$. If the test is satisfied, the program should output y, and otherwise it should continue looking.

Definition. Call a function f *computable* if it meets any of the conditions of Theorem 5.12.3.

Exercise 5.12.1. Write the IMP(L') program described above.

We know something about *semi*computable relations. What about the IMP(L) computable ones?

Proposition 5.12.4. *Let* $\mathbf{R}$ *be an* n-*ary relation on* L*. *And let* $\mathbf{R}^c$ *be the complement of* $\mathbf{R}$. *That is,* $\mathbf{R}^c$ *is the* n-*ary relation on* L* *whose members are those* n-*tuples that are not in* $\mathbf{R}$. *Then* $\mathbf{R}$ *is IMP*(L') *computable if and only if both* $\mathbf{R}$ *and* $\mathbf{R}^c$ *are generated by EFS*(**str**(L)) *procedures (i.e., if and only if both* $\mathbf{R}$ *and* $\mathbf{R}^c$ *are semicomputable).*

Exercise 5.12.2. Prove Proposition 5.12.4.

Exercise 5.12.3. Add to the machinery of IMP a PRINT instruction which, when executed, causes the value of the output identifier to be printed on some output device, but without halting the program. The addition of such an instruction makes it possible to write programs that output an infinite list of values rather than just one.

Let **R** be a relation on L^* and suppose **R** is the output of some EFS(**str**(L)) procedure. Show that some IMP(L') program using the additional PRINT instruction will print out exactly the members of **R**.

Exercise 5.12.4. Define a sequence $f_1, f_2, f_3, \ldots$ of binary functions on positive integers by the following conditions.

$f_1(x, y) = x + y$,

and for $n \geq 1$,

$$f_{n+1}(x, 1) = x$$
$$f_{n+1}(x, y+1) = f_n(x, f_{n+1}(x, y)).$$

(a) Prove $f_2(x, y) = x \cdot y$
(b) Prove $f_3(x, y) = x^y$
(c) Evaluate $f_4(2, 3)$ and $f_4(2, 4)$.

Exercise 5.12.5. (Exercise 5.12.4 continued). Define a three-argument function F on words as follows. If i, n, m are numbers ≥ 1 and $\bar{i}$, $\bar{n}$, $\bar{m}$ are their number representatives, then $F(\bar{i}, \bar{n}, \bar{m})$ is the number representative for $f_i(n, m)$. And on all other arguments the value is $\bar{0}$. Prove F is computable.

The function F in the exercise above is a close relative of what is known as the *Ackermann function*. It can be shown that it is *not* computable in $\text{IMP}^-(L)$, though for each particular $i \geq 1$, the 2 argument function $F(\bar{i}, x, y)$ is. We do not show this here as the proof is lengthy. But the result, combined with Exercise 5.12.5, implies that $\text{IMP}^-(L)$ is computationally weaker than IMP(L). All computation can be done in IMP(L) using counting loops and, at most, one while loop. Ackermann's function establishes that one cannot do without that while loop.

Exercise 5.12.6. Let $L = \{1\}$, so that the members of L^* are exactly the number representatives. For simplicity we will think of them as being numbers. Let **R** be an infinite subset of L^* whose characteristic function $c_{\mathbf{R}}$ is IMP(L') computable. Write an IMP(L') program to compute a function f such that:

(a) The domain of f is all of L^*(f is total).
(b) f is increasing, in that $n < m$ implies that $f(n) < f(m)$.
(c) The range of f is **R**.

Exercise 5.12.7. (converse of Exercise 5.12.6) Again let $L = \{1\}$. Suppose there is an IMP(L') computable function f such that

(a) The domain of f is all of L^*.
(b) f is increasing, $n < m$ implies $f(n) < f(m)$.
(c) The range of f is **R**, an infinite set.

Show that the charactersitic function $c_{\mathbf{R}}$ of **R** is IMP(L') computable.

When going from EFS to IMP, the alphabet expanded from L to L'. This is not necessary. The kinds of coding techniques discussed in Chapter 4, from larger alphabets into smaller ones, all can be applied to IMP and LOG style languages as well as to EFS style ones. The consequence is, in every result stated in this section, all mention of L' could be replaced by mention of L. We do not present a proof of this as we have no need of the result. You might wish to attempt a proof yourself, however.

5.13 THE CHURCH-TURING THESIS

There are real-world notions of computation and algorithm, both for people and for machines. We have created a mathematical model of those notions, using EFS(**str**(L)). How good is the model? How well does it capture the ideas we want to investigate? This question itself is not subject to mathematical investigation. For instance, suppose one mathematically models the behavior of a vibrating string by writing a certain differential equation. One cannot rigorously prove the equation accurately reflects the behavior of real vibrating strings, for a string is not a mathematical object that can be the subject of a mathematical argument. Instead, one can empirically confirm, with some degree of certainty, that the mathematical model is correct by comparing features of the model with the observed behavior of many real-world vibrating strings. Eventually one may develop confidence in the model, and use it to make predictions about observed strings.

We are in a similar position. We have a mathematical model of computation and of rule for computing, using EFS(**str**(L)). We must verify that the model is a good one. To a large extent that is what the present and the previous chapters have been about. The language IMP(L) is like computing languages in widespread use. We have shown the equivalence of IMP(L) and EFS(**str**(L)). We present this as evidence that the EFS model is a correct one. Implementability on register machines is another piece of evidence. Further, a variety of data structures are used in computational work and in the previous chapter we showed many of these were available in EFS(**str**(L)); others could easily be added to the list. We present this as further justification for our model.

The issue of a mathematical model for computation was first raised in the 1930s. Alonzo Church created a mathematical construct called the *lambda calculus* and, after much investigation he asserted in 1936 that it was an accurate mathematical model of the real-world idea of computation (the first such one). His thesis was that the functions one would intuitively call computable are exactly those representable, in a technical sense, in the lambda calculus. Further, in 1936, Alan Turing published a simple mathematical model of a computing machine, and asserted that this was an accurate model: A function is intuitively computable just in case it is one for which a suitable Turing machine can be defined.

The lambda calculus and Turing machines are mathematical creations that seem to have no connection with each other. They are strikingly different. But it was soon proved that the two were computationally equivalent (and here we mean mathematically proved; we are dealing with proper mathematical concepts now). Turing's model and Church's model were intertranslatable. What is now known as the *Church-Turing Thesis* is that there is a real notion of computability, which we have some intuitive knowledge of, and their models correctly capture this notion.

Over the years many other mathematical constructs were proposed. A few are: Kleene's mu-calculus (1936), Gödel-Herbrand equations (1936), Post's normal systems (1943), Markov algorithms (1954), Smullyan's elementary formal systems (1961), and Shepherdson-Sturgis register machines (1963). Each one was created after an analysis of a different aspect of real-world computation. All these attempts, and others such as EFS(**str**(L)) and IMP(L), turned out to be equivalent. This is strong evidence that a correct mathematical model has been constructed. Thoughtful analyses from a variety of viewpoints have all led to (equivalent versions of) the same model.

Two items from the list above should be singled out here. Register machines, seen in this chapter, are mathematical models of computers, as are Turing machines. And elementary formal systems are, essentially, the EFS(**str**(L)) languages we have been using.

Though it may seem paradoxical, the human race got along for much of its history without a mathematical model of computability simply because its interests were in things that *could* be computed. If one asserts something can be computed, the easiest way to prove it is to say how. And for this, no general theory of computability is needed. Most people recognize an algorithm when they see one. But in the twentieth century people became interested in the *impossibility* of computation for certain problems (we will see examples in the next chapter). It is for such issues that a mathematical model of computability becomes essential. In order to show computation is possible, we give an algorithm. In order to show it is not, we must show no algorithm is suitable, which means we need a satisfactory definition of algorithm. It is no coincidence that both Church and Turing presented their models in papers establishing computational impossibilities.

The Church-Turing Thesis is almost universally accepted by people in this field. An equivalent version of it is that a function is intuitively computable just when its graph is generated by some EFS procedure. There is a similar assertion for relations. In the rest of this book we take it for granted that, if we show no EFS procedure exists for a certain task, then that task is not subject to computational methods in the intuitive, everyday sense. In investigating our mathematical model we believe we are discovering things about what can and can't be done in the real world.

5.14 EFFICIENCY

Some programs are more efficient than others at the same task. *Computational complexity theory* is the study of the "cost" of running a program, particularly of its time and space requirements. It is a subject we have been ignoring, but now we have a good opportunity to discuss time requirements a little, and to state two fundamental open problems of computer science. If we make the simplifying assumption that each register machine instruction takes one unit of time to execute, then trace length becomes a measure of time.

Definition. Suppose that **R** is a set of words over L, and the characteristic function $c_{\mathbf{R}}$ of **R** is computable, say by the register machine module M. The *time function* for M is defined by $T_M(w) =$ number of lines of the trace for M using w as input.

Since M computes a characteristic function, which is defined on all inputs, trace length is always finite, so the definition of T_M is meaningful. But in general, time functions are much too irregular to be of any use. Things can be smoothed out by considering upper bounds instead. These are allowed to be bigger than absolutely necessary so that, perhaps, they will be easier to describe mathematically. Also, though it is of lesser importance, it is convenient to make them depend only on word lengths, and not on the words themselves.

Definition. For a word w over an alphabet L, $|w|$ is the length of w.

Definition. Let f be a function from numbers to numbers. f is a *time bound* for module M if, for each word w, the number of lines of the trace of M using w as input is $\leq f(|w|)$. In other words, f is a time bound for M if $T_M(w) \leq f(|w|)$. M is said to be *polynomial time* if it has a time bound that is a polynomial. Similarly, M is said to be *exponential* time if it has a time bound that is an exponential function (but none that are polynomial).

If we have two programs for the characteristic function of **R** and one is polynomial time while the other is exponential time, we would generally be better off with the polynomial time program. Likewise, between two polynomial time programs, the degrees of the polynomials are important; the smaller the degree the better.

Definition. A set of **R** of words is said to be *recognizable in deterministic polynomial time* if at least one register machine module for its characteristic function is polynomial time. Sets meeting this condition are said to be in the *P class,* for (deterministic) polynomial time. A similar definition of a set being recognizable in deterministic exponential time can be given.

We have been measuring time requirements using register machine programs, which are deterministic. We also want to use EFS(**str**(L)) procedures, which are nondeterministic. But nondeterminism means that traces are not unique. For instance, if the rules for constructing a trace allow us to add a certain line, we are also allowed to add it several times, creating a superficially different trace. There are other less obvious ways different EFS traces can arise that accomplish the same thing. So, we single out the "best" candidate among EFS traces, which we take to be the shortest. We measure EFS trace length not by counting lines but by counting symbols, because single lines can be arbitrarily long (to create a substitution instance of a procedure statement, words of *any* length can be substituted for variables).

Definition. Once again, suppose that **R** is a set of words with a computable characteristic function. It follows by Proposition 5.12.4 that some EFS(**str**(L)) procedure has the set **R** as output, say R_PROCEDURE. f is a *time bound* for R_PROCEDURE provided, for each word $w \in \mathbf{R}$, the shortest trace for R_PROCEDURE that ends with R_PROCEDURE(w) has $\leq f(|w|)$ symbols.

It follows from this definition that if no trace with $\leq f(|w|)$ symbols ends in R_PROCEDURE(w), then $w \notin \mathbf{R}$. Again, polynomial, exponential, etc. distinctions are made, as above. Finally, these notions too are transferred from procedures to sets by considering "best" procedures.

Definition. The set **R** is said to be *recognizable in non-deterministic polynomial time,* and to be in the *class NP,* if at least one EFS(**str**(L)) procedure for it has a polynomial as a time bound.

The definitions we gave of P and of NP depend closely on details of the programming languages developed in this book. But as a matter of fact, the classes P and NP themselves are quite *robust* in the sense that attempts to define comparable notions using almost any deterministic and nondeterministic programming languages will lead to the same classes. For the deterministic version we could have counted the number of steps of a register machine when executing an IMP program for the characteristic function. Other variations can easily be imagined. Commonly, Turing machines are used, in both deterministic and nondeterministic versions. Thus, the classes P and NP have a "real" existence, and are not programming language artifacts. The same can not be said for certain finer classifications. For instance, a set **R** may be recognizable in polynomial time using programs written in two different deterministic programming languages, but the *degrees* of the polynomials involved may be different. It is the broader classification of being in P that will not change.

Sets that are recognizable in deterministic exponential time, but no better, are useless, practically. Computation time grows so quickly with input that any reasonable real-world time limitation is soon exceeded. Even sets that are recognizable in deterministic polynomial time may be pretty bad if the polynomial is of high degree. But as a first approximation, sets in the P class are those we can plausibly think about programming usefully. Such sets are sometimes called *tractable.*

On the nondeterministic side, many sets that people are interested in are in the NP class. The first example known was the set **R** of all *satisfiable* formulas of propositional logic (Boolean algebra). A formula, built up from atomic formulas using $\wedge$, $\vee$, and $\neg$ (negation), is satisfiable if some assignment of truth values (true and false) to its atomic parts will cause the whole formula to evaluate to true. An efficient program to recognize this set would be of great use in logic programming, in meeting constraints on data bases, and in simplifying circuit diagrams, which can be modeled by logic expressions. It is not difficult to verify that the set **R** is in the class *NP,* but it would take us too far afield to do so here.

Another example is the so-called traveling salesperson problem. Given a list of cities, and distances between them, what is the shortest path that visits each city once and returns to its starting point? This can be phrased in terms of words by thinking of the list of cities and distances as a single big word. Now consider the 2-place relation **R** such that $\langle u, v\rangle \in \mathbf{R}$ provided u is a list of cities and distances, and v is a number (representative) such that some trip visiting each city once and returning home has length $\leq v$. We defined *P* and *NP* for sets, but the definitions could be given just as well for relations. Then **R** turns out to be in the *NP* class. The traveling salesperson problem has obvious applications in scheduling.

The examples above are two of a very long list of known *NP complete* problems. The qualifier "complete" means that, in a certain technical sense, every other *NP* problem can be reduced to them. An "efficient" program for any one of them would allow us to write "efficient" programs for all of them. An example of something that is in the class *NP* but that is not known to be complete is the set of numbers (or number representatives) that are not prime.

Obviously an important question is: What is the relationship between *P* and *NP,* since *P* contains sets we can work with "efficiently," while *NP* contains many sets of known importance. We have given translation procedures between register machine modules and EFS(**str**(L)); what can we learn from them?

In one direction things are simple. A close analysis of the translation we gave from register machines to EFS(**str**(L)) shows that it turns polynomial time modules into polynomial time procedures. We do not go into details here, but an obvious consequence is $P \subseteq NP$.

The other direction is more critical. The heart of the matter is the translation we gave from EFS(**str**(L)) into IMP(L') in §5.12. Since the translation from IMP(L') to register machine will only add steps, making things worse instead of better, we concentrate on asking how bad is the translation we gave from EFS(**str**(L)) to IMP(L'). Most of the details can safely be ignored. The key item is the program given in §5.12 under the name S_PROCEDURE_SEMI_CHAR. Recall the idea behind that program. It searches for a suitable EFS trace by trying word after word until it finds the right one. Further, it conducts that search by first trying all words of length 2, then all of length 3, then 4, and so on. If the alphabet L' has p letters, there will be p^2 words of length 2, p^3 words of length 3, and so on. If the shortest EFS trace having the desired last line contains k symbols, to find it by the method just described will require us to first examine and discard $p^2 + p^3 + \ldots + p^{k-1}$ words. This sums to $(p^k - p^2)/(p-1)$, which behaves like p^k as k grows. In other words, if the shortest appropriate EFS procedure trace has k symbols, the IMP translate must execute on the order of p^k steps. If the EFS procedure is polynomial time, its translate will be exponential time.

Thus, we have failed to show that $NP \subseteq P$. This does not mean NP is not a subset of P. It merely means our translation procedure is not efficient enough to establish that it is. After all, **R** is in the P class if *some* register machine module for **R** is polynomial time. It is conceivable that a better translation procedure could produce one. But, as a matter of fact, the question of whether $NP = P$ or $NP \neq P$ is a major open problem. The general belief is that P and NP are different. In effect, this would say that no implementation of EFS on register machines can be significantly better than the rather crude one we gave.

Now, having presented the problem, we must leave it. A detailed discussion of the work towards a solution done thus far would take us beyond the bounds of this book. Instead we briefly present another major open problem in computational complexity, and then we return to our earlier practice of ignoring efficiency issues entirely.

The definition of deterministic polynomial time used characteristic functions. A program that runs in polynomial time and computes a characteristic function answers questions about what is *not* in the set **R** quite as efficiently as it answers questions about what *is* in the set. It follows that if **R** is in the class P, so is its complement, $\mathbf{R}^c$. In other words, if we let co-P be the class of all sets whose complement is in P, then $P = \text{co-}P$. But this symmetry does not seem to carry over to NP and co-NP (defined to be the class of sets **R** whose complement, $\mathbf{R}^c$, is in NP). The problem is this. Suppose we have an EFS(**str**(L)) procedure R_PROCEDURE to generate **R**, and it has a polynomial time bound f. Suppose we suspect $w \notin \mathbf{R}$; how would we use R_PROCEDURE to confirm our suspicion? Well, if w were in **R**, some trace with $\leq f(|w|)$ symbols would tell us so, and there are only a finite number of traces with $\leq f(|w|)$ symbols. Check them all out, and if none of them says $w \in \mathbf{R}$, it must be that $w \notin \mathbf{R}$. The trouble is, as we saw earlier, if we add up the lengths of all candidates for EFS traces that have $\leq f(|w|)$ symbols we get a number that grows exponentially fast with $|w|$. Thus, although we have a way of using R_PROCEDURE to get at $\mathbf{R}^c$, the method requires exponential, not polynomial time.

We have not shown that $NP \neq \text{co-}NP$; we have only seen that the most obvious thing does not work. After all, if $\mathbf{R} \in NP$ because R_PROCEDURE generates it in polynomial time, it may still be that $\mathbf{R}^c \in NP$ because it is generated by some procedure in polynomial time, but a procedure having no obvious connection with R_PROCEDURE at all. The question of whether $NP = \text{co-}NP$ or $NP \neq \text{co-}NP$ is another one of the major open problems in computer science. Obviously, if it turned out that $P = NP$, then NP would equal co-NP since $P = \text{co-}P$. But not a great deal more is known about answers to these questions, though a great deal is known about what will not settle them.

5.15 BACKGROUND

In the 1930s a family of functions on the natural numbers was defined, and given the name of *recursive functions*. Actually, equivalent versions were defined independently by several people, though it was not immediately obvious that the various definitions characterized the same things. But it was soon established. What is now known as the *Church-Turing Thesis* is that: the class of recursive functions is the formal counterpart of the class of functions on numbers that we would intuitively call computable. [Rogers 1967] is a thorough presentation of the mathematical properties of the class of recursive functions.

In terms of the present book, the class of recursive functions can be characterized in several ways. Let f be a (perhaps partial) function from $\{0, 1, 2, \ldots\}$ to itself. And let f' be the corresponding function on the set of words over the alphabet $\{1\}$, taking the number representative for n to the number representative for $f(n)$. Then, f is a (partial) recursive function if any of the following equivalent items holds:

1. The graph of f is generated by some procedure of EFS(**integer**);
2. There is an operator Φ defined by a procedure of EFS(**integer**) such that

$$\Phi(\{n\}) = \begin{cases} \{f(n)\} & \text{if } f \text{ is defined on } n \\ \varnothing & \text{otherwise;} \end{cases}$$

3. f' is computable in our sense, that is, it meets any of the conditions of Theorem 5.12.3.

The list of equivalences can easily be lengthened. For instance, we could use a string language with an alphabet $\{0, 1\}$, and use base 2 names for numbers instead of naming them with strings of 1's. The obvious variations are still equivalences.

For most of this book we have been working with relations, not functions. For a recursion theorist, a relation generated by a procedure of EFS(**integer**) is called *recursively enumerable*. For a suggestion of why this name is appropriate, see Exercise 5.12.3.

The first formal definition of the class of functions now called recursive was in [Church 1936]. It was phrased in terms of the *lambda calculus*. An equivalent definition of recursive function was given in [Turing 1936], and independently in [Post 1936], in terms of mathematically idealized computing devices now known as *Turing machines*. Incidentally, these mathematical idealizations were created before the real machines we think of them as modeling. A direct proof that Turing machines can be simulated using a logic programming language (essentially EFS(**log**(**S**)) from Chapter 2, §2.10) can be found in [Tärnlund 1977].

Yet another equivalent characterization is due to Kleene, [Kleene 1936] and is fully developed by him in [Kleene 1952], where equivalence proofs between it and other characterizations are given. Loosely, the idea is to start with a simple class of functions, called *primitive recursive*, and then add an unbounded search operator. The study of the class of primitive recursive functions (though not the name) goes back to [Dedekind 1888]. In terms of the machinery of the present book, a function f on numbers is primitive recursive if the related function f' on number representatives is $\text{IMP}^-(\{1\})$ computable. This is not Kleene's definition, but it is provably equivalent. The primitive recursive functions constitute an important subclass of the class of all recursive functions and much is known about them. They fall naturally into a hierarchy of difficulty, where the computational complexity of f can be measured by the number of nested counting loops necessary in an $\text{IMP}^{-}(\{1\})$ program for f'. See [Meyer and Ritchie 1967].

It was essentially shown in §5.12 that all programming can be done using only the structured machinery of IMP(L) and with at most one while loop. Basically this is Kleene's normal form theorem [Kleene 1936]. There is a direct proof of this for flowchart programs in [Böhm and Jacopini 1966]. And there is an improved version in [Ashcroft and Manna 1972] in which program "topology" is preserved during elimination of GOTO's.

Register machines come from [Shepherdson and Sturgis 1963]. As originally presented, they manipulated numbers not words; the version presented here is a straightforward modification.

Yet another characterization of the recursive functions uses Post productions [Post 1943]. Post's machinery was the direct ancestor of elementary formal systems as found in [Smullyan 1961] and is thus the indirect ancestor of EFS(**str**(L)).

While recursion theory began as a theory of numbers, extending it to words was an early development. One can do this indirectly using codings (Gödel numberings) or directly, as we have done here. Turing machines, in fact, work with words directly, and only indirectly with numbers. The same is true of Post productions. It has proved of use to see what can be done using Turing machines or Post productions that are limited in some way. This is part of the subject of *automata theory*. It is not something we go into here, except to point out that the automata theoretic notion of being *strictly rudimentary* coincides with what we have called $\text{LOG}^{-}(L)$ definable. The notion of strict rudimentary comes from [Smullyan 1961].

Computational complexity issues play an important role in both theoretical and applied computer science. The class *NP* was first investigated in [Cook 1971], and the idea of *NP* completeness developed in [Karp 1972]. By now, a long and still growing list of *NP* complete problems is known.

REFERENCES

[Ashcroft and Manna 1972] The translation of "go to" programs to "while" programs, E. Ashcroft and Z. Manna, *Proc. IFIP Congress 1971,* North Holland Publishing Co., Amsterdam, pp. 250–255, 1972.

[Böhm and Jacopini 1966] Flow diagrams, Turing machines and languages with only two formation rules, C. Böhm and G. Jacopini, *Communications of the Association for Computing Machinery,* Vol. 9, pp. 366–371, 1966.

[Church 1936] An unsolvable problem of elementary number theory, A. Church, *The American Journal of Mathematics,* Vol. 58, pp. 345–363, 1936, reprinted in *The Undecidable,* M. Davis, editor, Raven Press, Hewlett, N. Y., pp. 88–107, 1965.

[Cook 1971] The complexity of theorem proving procedures, S. Cook, *Proc. Third Annual Association for Computing Machinery Symposium on the Theory of Computing,* pp. 151–158, 1971.

[Dedekind 1888] *Was sind und was sollen die Zahlen?,* R. Dedekind, Braunschweig, 1888, English translation by Beman as The nature and meaning of numbers, in *Essays on the Theory of Numbers,* Open Court, Chicago, 1901, reprinted by Dover Publications, 1963.

[Karp 1972] Reducibility among combinatorial problems, R. Karp, in *Complexity of Computer Computations,* R. Miller and J. Thatcher, editors, Plenum Press, New York, pp. 85–103, 1972.

[Kleene 1936] General recursive functions of natural numbers, S. Kleene, *Mathematische Annalen,* Band 112, Heft 5, pp. 727–742, 1936, reprinted in *The Undecidable,* M. Davis, editor, Raven Press, Hewlett, N. Y., pp. 236–253, 1965.

[Kleene 1952] *Introduction to Metamathematics,* S. Kleene, D. Van Nostrand, Princeton N. J., 1952.

[Meyer and Ritchie 1967] The complexity of loop programs, A. Meyer and D. Ritchie, *Proc. of the 22nd National Conference, Association for Computing Machinery,* pp. 465–469, 1967.

[Post 1936] Finite combinatory processes—Formulation I, E. Post, in *Journal of Symbolic Logic,* Vol. 1, pp. 103–105, 1936, reprinted in *The Undecidable,* M. Davis, editor, Raven Press, Hewlett, N. Y., pp. 288–291, 1965.

[Post 1943] Formal reductions of the general combinatorial decision problem, E. Post, *American Journal of Mathematics,* Vol. 65, pp. 197–215, 1943.

[Rogers 1967] *Theory of Recursive Functions and Effective Computability,* H. Rogers, Jr., McGraw-Hill, New York, 1967.

[Shepherdson and Sturgis 1963] Computability of recursive functions, J. Shepherdson and H. Sturgis, *Journal of the Association for Computing Machinery,* Vol. 10, pp. 217–255, 1963.

[Smullyan 1961] *Theory of Formal Systems,* revised edition, R. Smullyan, Princeton University Press, Princeton, N. J., 1961.

[Tärnlund 1977] Horn clause computability, S. Tärnlund, *Nordisk Tidskrift for Informationsbehandling (BIT),* Vol. 17, pp. 215–226, 1977.

[Turing 1936] On computable numbers, with an application to the entscheidungsproblem, A. Turing, *Proc. London Mathematical Society,* Ser. 2, Vol. 42, pp. 230–265, 1936–1937, corrections ibid., Vol. 43, pp. 544–546, 1937. Reprinted in *The Undecidable,* M. Davis, editor, Raven Press, Hewlett, N. Y., pp. 116–154, 1965.

6
PROGRAMS AS DATA

6.1 INTRODUCTION

In Chapter 5, we showed how to turn EFS procedures into register machine programs. Now we write an interpreter for EFS in EFS itself! To be more precise, consider the 2-place relation **R** between procedures and words: w is an output of the EFS(**str**(L)) procedure PROC. A procedure output w is a word over the alphabet L, and PROC itself is a word over a larger alphabet L'. So relation **R** is really a relation on words. We write an EFS(**str**(L')) procedure UNIV whose output is this very relation **R**. Such a procedure is "universal" in the sense that it tells us about the output of *all* EFS(**str**(L)) procedures. And writing the UNIV procedure is straightforward. We just take the rules governing the construction and behavior of EFS procedures from Chapter 1 and rewrite them as EFS procedures.

Just as in Chapter 5, we use a larger alphabet L' containing "stand-ins" for the various punctuation symbols and identifier letters we need to write EFS(**str**(L)) procedures. But by applying results from Chapter 4 on implementing **str**(L') in **str**(L) we show there is a single generated relation of EFS(**str**(L)) itself from which all other generated relations can be easily obtained, a *universal generated relation.*

The existence of a universal relation is an essential tool that will allow us to establish major negative results. We show there are many important properties of EFS procedures, and IMP programs too, that are simply not computable. One cannot write EFS procedures, or IMP programs, to decide when those properties hold. A better statement must wait until later in the chapter, but then we will establish some fundamental limitations on the abilities of computers.

6.2 THE ALPHABET

Procedures of EFS(**str**(L)) are written using an alphabet containing arrows and commas and the like. We can not make these extra symbols into official letters because ambiguity would result so, as in Chapter 5, §5.10, we introduce "stand-ins" for them. In that chapter we looked at traces, which are variable free, but not at procedures themselves, as we will do here. This accounts for the two extra symbols used here but not in Chapter 5.

Definition. E is the alphabet consisting of $\Rightarrow$, $_0$, $\llbracket$, $\rrbracket$, $\mathbb{A}, \ldots, \mathbb{Z}$, $_=$, #, $*$, and !. And for an alphabet L, by L' we mean $L \cup E$.

The intention is that we use

$\Rightarrow$	in place of	$\rightarrow$
$_0$	in place of	,
$\llbracket, \rrbracket$	in place of	(,)
$\mathbb{A}, \ldots, \mathbb{Z}, _=$	in place of	$A, \ldots, Z, \_$

We use # to play the role of the colon, semicolon and period in procedures, and also of newline in traces. Finally, EFS procedures involve variables, say v', v'', v''', . . . is the official list (we have been using x, y, z, . . . for them, on an informal basis). The aliases for these variables will be $*!*$, $*!!*$, $*!!!*$, ...

For example, say that S is the *procedure statement* CON(v', ab, v'')$\rightarrow$ OUT(v'') where a, $b \in L$. Then its alias is the *word* T written over the alphabet L': $\mathbb{CON}\llbracket *!*_0 ab_0 *!!* \rrbracket \Rightarrow \mathbb{OUT}\llbracket *!!* \rrbracket$. Further, suppose we have the following procedure of EFS(**str**(L)).

NAME(n):
    S_1;
    S_2;
    ⋮
    S_k.

The alias for this is the following word over L', in which for each procedure statement S_i, T_i is its word alias defined above: $\mathbb{NAME}\llbracket \bar{n} \rrbracket \# T_1 \# T_2 \# \ldots \# T_k \#$.

We can think of each procedure of EFS(**str**(L)) as being represented by, or more loosely, as *being* a word over L'. Then the relationship between a procedure of EFS(**str**(L)) and its outputs is actually a relation on L'^*. It is this relation we will show is generated in EFS(**str**(L')).

6.3 ELEMENTARY SYNTAX

We show the words over L' that are aliases for EFS(**str**(L)) procedures in the basic work space constitute a generated set. To do this, we write formal counterparts of the definitions of Chapter 1, §1.3. We do this in detail because of the importance of the final result. In working with EFS(**str**(L')) we freely use procedures written earlier in this book. In particular, in Chapter 5, §5.9, 5.10 we gave several definitions in LOG(L), and showed they corresponded to EFS procedures. We use a work space containing all these defined relations. Now, the following are generated relations of EFS(**str**(L')).

1. *LETTER_OF_L*. The alphabet L is a subalphabet of L'. Since it is finite, it is trivially generated.
2. *WORD_OVER_L*. Words over L are needed in substitution instances of procedure statements. See Exercise 4.5.1.

3. *CON,* where $\text{CON}(x_1, \ldots, x_n, y)$ means that y is the concatenation of $x_1, \ldots, x_n$. This is really a family of relations. See Chapter 5, §5.10.

4. *VARIABLE.* Recall that our choice of alias for variables is words of the form ∗!!...!∗.

   VARIABLE (1):
   EXCLAIM_STRING(!);
   EXCLAIM_STRING(x)→CON(x, !, y)→EXCLAIM_STRING(y);
   EXCLAIM_STRING(x)→CON(∗, x, ∗, y)→VARIABLE(y).

5. *BEGINS, NOT_BEGINS, ENDS, NOT_ENDS, SUBWORD, NOT_SUBWORD.* See Chapter 5, §5.9, including Exercise 5.9.4.

6. *IDENTIFIER_LETTER,* the alphabet 𝔸, ..., ℤ, ₌. This is finite, hence it is generated.

7. *IDENTIFIER.* An identifier is a nonempty word made up of 𝔸, ..., ℤ, ₌ that does not begin or end with ₌.

   IDENTIFIER (1):
   IDENTIFIER_LETTER(x)→WORD(x);
   WORD(x)→IDENTIFIER_LETTER(y)→CON(x, y, z)→WORD(z);
   WORD(x)→NOT_BEGINS(₌, x)→NOT_ENDS(₌, x)→
   IDENTIFIER(x).

8. *TERM.* A term is a variable or a word over L. Generated relations are closed under ∨, and we have (2) and (4).

9. *ATOMIC.* An atomic statement is an identifier followed by a list of terms separated by commas and enclosed in parentheses.

   ATOMIC (1):
   TERM(x)→TERM_LIST(x);
   TERM_LIST(x)→TERM(y)→CON(x, ₒ, y, z)→TERM_LIST(z);
   IDENTIFIER(x)→TERM_LIST(y)→CON(x, ⟦, y, ⟧, z)→
   ATOMIC(z).

10. *UNRES_ATOMIC.* In the basic work space, only CON is reserved. The basic work space is enough to consider, by the discussion in Chapter 1, §1.5.

    UNRES_ATOMIC (1):
    ATOMIC(x)→NOT_BEGINS(ℂ𝕆ℕ⟦, x)→UNRES_ATOMIC(x).

11. *STATEMENT.* A procedure statement is a list of atomic statements separated by arrows.

    STATEMENT (1):
    ATOMIC(x)→STATEMENT(x);
    STATEMENT(x)→ATOMIC(y)→CON(x, ⇛, y, z)→
    STATEMENT(z).

12. *ACCEPTABLE_STATEMENT.* A procedure statement is acceptable if CON is not used in the assignment position.

ACCEPTABLE_STATEMENT (1):
UNRES_ATOMIC(x)→ACCEPTABLE_STATEMENT(x);
STATEMENT(x)→UNRES_ATOMIC(y)→CON(⇛, y, z)→
→ENDS(z, x)→ACCEPTABLE_STATEMENT(x).

13. *HEADER.* A header is an identifier followed by a number (representative) in parentheses. We assume a NUMBER_REP procedure has already been written.

HEADER (1):
IDENTIFIER(x)→NUMBER_REP(y)→CON(x, ⟦, y, ⟧, z)
→HEADER(z).

14. *PROCEDURE.* A procedure consists of a header followed by a list (possibly empty) of acceptable statements, all separated by suitable punctuation, # in the alphabet L'.

PROCEDURE (1):
HEADER(x)→CON(x, #, y)→PROCEDURE(y);
PROCEDURE(x)→ACCEPTABLE_STATEMENT(y)
→CON(x, y, #, z)→PROCEDURE(z).

We have now written an EFS syntax checker. PROCEDURE verifies that a string of symbols constitutes a correctly written procedure. A NOT_PROCEDURE procedure could also be written, but its analysis is more difficult and we will not need it. Syntax checkers can be written for all computer languages, of course. It is a routine, and elementary, part of compilers and interpreters. Devising algorithms for checking correct syntax becomes simpler with some knowledge of automata theory, a subject we do not go into here. In the following exercises, you are asked to write an EFS syntax checker for IMP(L). To begin, pick an alphabet L'' that is to play the role for IMP(L) that L' has been playing above. That is, L'' should extend L and contain enough extra letters to function as aliases for the special symbols of IMP(L). We leave this to you. L'' in the exercises refers to this.

Exercise 6.3.1. Write EFS(**str**(L'')) procedures for the following, using Chapter 5, §5.6 for guidance.

(a) IMP_TERM
(b) CONDITION
(c) ASSIGNMENT_STATEMENT
(d) STATEMENT.

Exercise 6.3.2. Write an EFS(**str**(L'')) procedure for IMP_PROGRAM.

6.4 PROCEDURE EXECUTION

We continue the sequence of EFS(**str**(L')) procedures begun in §6.3, writing a procedure for EFS(**str**(L)) trace, from which procedure output relations are easily obtained. To begin, we need the notion of substituting a constant (word of L^*) for a variable in a procedure statement. There are several ways this can be done. We parallel the construction of statement and statement-with-the-substitution-carried-out, beginning with terms and working upwards.

15. *TERM_SUB*(t, x, y, z). This is intended to take care of substitution in single terms: t is a term, x is a variable, y is a word over L, and z is the result of replacing x by y in t, which means $z = y$ if $t = x$, and $z = t$ if $t \neq x$.

TERM_SUB (4):
TERM(t)→VARIABLE(x)→WORD_OVER_L(y)
→EQUAL(x, t)→TERM_SUB(t, x, y, y);
TERM(t)→VARIABLE(x)→WORD_OVER_L(y)
→NOT_EQUAL(x, t)→TERM_SUB(t, x, y, t).

16. *TERM_LIST_SUB*(t, x, y, z). This is intended to represent that t is a list of terms (separated by commas), x is a variable, y is a word over L, and z is the result of replacing all occurrences of x by occurrences of y.

TERM_LIST_SUB (4):
TERM_SUB(t, x, y, z)→TERM_LIST_SUB(t, x, y, z);
TERM_LIST_SUB(q, x, y, r)→TERM_SUB(u, x, y, v)
→CON(q, $_0$, u, t)→CON(r, $_0$, v, z)
→TERM_LIST_SUB(t, x, y, z).

17. *SUB*(x, y, z, w). This is intended to represent that x is a procedure statement, y is a variable, z is a word over L, and w is the result of replacing all occurrences of y in x by occurrences of z.

SUB (4):
ATOMIC(x)→IDENTIFIER(u)→CON(u, ⟦, q, ⟧, x)
→TERM_LIST_SUB(q, y, z, r)→CON(u, ⟦, r, ⟧, w)
→SUB(x, y, z, w);
STATEMENT(u)→ATOMIC(v)→CON(u, ⇛, v, x)→SUB(u, y, z, q)
→SUB(v, y, z, r)→CON(q, ⇛, r, w)→SUB(x, y, z, w).

18. *PARTIAL_INSTANCE_OF*(x, y). This is intended to represent that y is a procedure statement and x is a partial substitution instance of y, that is, some variables of y have been replaced by words over L, not necessarily all of them.

PARTIAL_INSTANCE_OF (2):
STATEMENT(y)→PARTIAL_INSTANCE_OF(y, y);
PARTIAL_INSTANCE_OF(z, y)→SUB(z, u, v, x)
→PARTIAL_INSTANCE_OF(x, y).

19. *NOT_VARIABLE,* representing the set of words that are not variables. A word is not a variable (representative) if it does not begin with ∗, does not end with ∗, or begins and ends with ∗ but doesn't have a string of ! between. In the first lines below we enumerate all the letters of L' other than !.

NOT_VARIABLE (1):
    NOT_EXCLAIM(*a*);
    ⋮
    NOT_EXCLAIM(ℤ);
    NOT_EXCLAIM(*x*) → SUBWORD(*x*, *y*)
        → NOT_EXCLAIM_STRING(*y*);
    NOT_EXCLAIM_STRING(Δ);
    NOT_BEGINS(∗, *x*) → NOT_VARIABLE(*x*);
    NOT_ENDS(∗, *x*) → NOT_VARIABLE(*x*);
    NOT_EXCLAIM_STRING(*y*) → CON(∗, *y*, ∗, *x*)
        → NOT_VARIABLE(*x*).

20. *SUB_INSTANCE_OF*(x, y). This represents the relation that x is a substitution instance of y; *all* variables have been replaced by constants. The easiest way to say no variables are left is to say, whatever we see, it is not a variable. It is more convenient for us to give a LOG(L') definition here than an EFS(**str**(L')) procedure.

SUB_INSTANCE_OF(*x*, *y*) ⇔
        STATEMENT(*y*) ∧ PARTIAL_INSTANCE_OF(*x*, *y*)
        ∧ (∀*z* SUBWORD *x*)NOT_VARIABLE(*z*).

21. *LINE_OF.* This relation treats a word occurring between consecutive # occurrences as a line. Recall that # plays the role of semicolon, colon, and period in procedures, and newline in traces.

LINE_OF (2):
    NOT_SUBWORD(#, *x*) → CON(#, *x*, #, *z*)
        → SUBWORD(*z*, *y*) → LINE_OF(*x*, *y*).

22. *PROC_INSTANCE_OF*(z, x). The relation is that z is a substitution instance of a statement of procedure x.

PROC_INSTANCE_OF (2):
    PROCEDURE(*x*) → LINE_OF(*y*, *x*) → SUB_INSTANCE_OF(*z*, *y*)
        → PROC_INSTANCE_OF(*z*, *x*).

23. *CON_INSTANCE,* representing true instances of the concatenation relation for L^*.

CON_INSTANCE (1):
    WORD_OVER_L(*z*) → CON(*x*, *y*, *z*) → CON(ℂ𝕆ℕ⟦, *x*, ₀, *y*, ₀, *z*, ⟧, *w*)
        → CON_INSTANCE(*w*).

24. *TRACE_OF*(x, y). This is intended to represent that y is a procedure and x is a restricted trace of it. We allow the empty trace, consisting of just #, for convenience.

TRACE_OF (2):
  PROCEDURE(y) → TRACE_OF(#, y);
  TRACE_OF(w, y) → CON_INSTANCE(z) → CON(w, z, #, x)
    → TRACE_OF(x, y);
  TRACE_OF(w, y) → PROC_INSTANCE_OF(z, y) → CON(w, z, #, x)
    → TRACE_OF(x, y);
  TRACE_OF(w, y) → LINE_OF(u, w) → LINE_OF(v, w) → ATOMIC(u)
    → CON(u, ⇒, z, v) → CON(w, z, #, x) → TRACE_OF(x, y).

25. *ARITY_OF*(x, y). This represents that y is a sequence of n words over L, separated by commas, and x is the number representative $\bar{n}$ for n.

ARITY_OF (2):
  WORD_OVER_L(y) → ARITY_OF($\bar{1}$, y);
  ARITY_OF(r, q) → WORD_OVER_L(s) → CON(q, $_0$, s, y)
    → CON(r, $\bar{1}$, x) → ARITY_OF(x, y).

26. *OUTPUT*(x, y, z). This represents that x is a procedure whose output is an n-place relation, y is $\bar{n}$, the representative for n, and z is an n-tuple in the output of x (written without enclosing parentheses).

OUTPUT (3):
  PROCEDURE(x) → BEGINS(u, x) → HEADER(u) → IDENTIFIER(i)
    → CON(i, ⟦, y, ⟧, u) → ARITY_OF(y, z) → CON(i, ⟦, z, ⟧, v)
    → TRACE_OF(t, x) → LINE_OF(v, t) → OUTPUT(x, y, z).

27. Finally, a *family* of relations, GENERATES_ONE, GENERATES_TWO, etc., one for each positive integer n. The nth one, GENERATES_N, is the $n+1$-place relation such that GENERATES_N(y, $x_1, \ldots, x_n$) if y is a procedure that outputs n-tuples, and $\langle x_1, \ldots, x_n \rangle$ is in the output of y.

GENERATES_N ($n + 1$):
  WORD_OVER_L(x_1) → WORD_OVER_L(x_2)
    → . . . → WORD_OVER_L(x_n)
    → CON(x_1, $_0$, x_2, $_0$, . . . , $_0$, x_n, z)
    → OUTPUT(y, $\bar{n}$, z) → GENERATES_N(y, $x_1, x_2, \ldots, x_n$).

Exercise 6.4.1. The idea behind the substitution procedure SUB above was to use the machinery by which statements are constructed from simpler statements. It is also possible to use a method that goes through a statement from left to right, replacing variable occurrences as encountered. Write an alternate SUB procedure based on this idea.

Exercise 6.4.2. Demonstrate, again, that the OUTPUT relation (item 26) is generated, but without using traces. Make use of Proposition 3.5.1 instead.

It is certainly possible to do for register machines what we have done for EFS(**str**(L)). That is, one can write an EFS(**str**(L'')) procedure whose output is the $n+2$-place relation consisting of all $\langle x, y_1, \ldots, y_n, z\rangle$ such that x is a module that accepts n inputs and which, when given $y_1, \ldots, y_n$ as inputs, produces z as output. To do this, the notion of register machine trace must be captured in a EFS procedure, much like we did with EFS traces above. We do not list this as an exercise. It is a project of some complexity. We recommend that you think about it seriously, whether or not you actually carry the project through to completion.

6.5 INTRODUCING CONSTANTS

Suppose we have an $n+m$-place relation R on L^* and $c_1, \ldots, c_n$ are n words of L^*, constants. Then we can define an m-place *section* relation S by $S(x_1, \ldots, x_m) \Leftrightarrow R(c_1, \ldots, c_n, x_1, \ldots, x_m)$. For instance, if R is the BEGINS relation, and we define $S(x) \Leftrightarrow \text{BEGINS}(c, x)$ then S consists of all words beginning with c. The relation S that we get depends on which constants $c_1, \ldots, c_n$ we use (and on R of course, but let's keep that fixed for now). In Chapter 2 we showed that if R is a generated relation then S will also be generated. We did this by saying how to turn a procedure for R into one for S. The construction went like this. Suppose we have a procedure for R, and S is an unreserved identifier that does not occur in the R procedure. Then we just write

S(m):
  (procedure body of R)
  $\text{R}(c_1, \ldots, c_n, x_1, \ldots, x_m) \rightarrow \text{S}(x_1, \ldots, x_m)$.

We have just defined an informal function: You tell me what $c_1, \ldots, c_n$ you have in mind, and I'll give you a procedure whose output is S, where $S(x_1, \ldots, x_m) \Leftrightarrow R(c_1, \ldots, c_n, x_1, \ldots, x_m)$. We now show a formal version of this function is computable in EFS(**str**(L')).

Proposition 6.5.1. *Let* R *be an EFS*(***str***(L)) *procedure whose output is an* n + m*-place relation. There is an* n *argument function* f *such that:*

1. *for* $\mathrm{c}_1, \ldots, \mathrm{c}_n \in \mathrm{L}^*$, $\mathrm{f}(\mathrm{c}_1, \ldots, \mathrm{c}_n)$ *is a procedure (thought of as a word over* L'*) for the relation* S, *where* $\mathrm{S}(\mathrm{x}_1, \ldots, \mathrm{x}_m) \Leftrightarrow \mathrm{R}(\mathrm{c}_1, \ldots, \mathrm{c}_n, \mathrm{x}_1, \ldots, \mathrm{x}_m)$, *and*
2. *the graph of* f *is generated in EFS*(***str***(L')).

Proof. We have a procedure R, which we can think of as being a certain word over L'. We write R^0 for this word, the L' alias for the *procedure* R. Also, R is an identifier, representing procedure output, and as such it has an alias in L'^*. We write $\mathbb{R}$ for the L' alias of the *identifier* R. Let S be an unreserved identifier not used in procedure R. S has an alias in L'^*; we write $\mathbb{S}$ for it.

Finally, we created variable aliases in L'^* (see item 4 in §6.3). For readability we write $\mathbb{U}_1$ for $*!*$, $\mathbb{U}_2$ for $*!!*$, etc. Now, the following procedure of EFS(**str**(L')) has as output the graph of the function we want.

$$
\begin{aligned}
&\mathrm{F}\ (n+1):\\
&\quad \mathrm{WORD\_OVER\_L}(z_1) \rightarrow \ldots \rightarrow \mathrm{WORD\_OVER\_L}(z_n)\\
&\qquad \rightarrow \mathrm{HEADER}(y) \rightarrow \mathrm{CON}(y, x, R^0)\\
&\qquad \rightarrow \mathrm{CON}(\mathbb{S}[\![\bar{m}]\!], x, \mathbb{R}[\![, z_1, {}_0, \ldots, {}_0, z_n, {}_0,\\
&\qquad\quad \mathbb{U}_1{}_0 \ldots {}_0\mathbb{U}_m]\!] \Rightarrow \mathbb{S}[\![\mathbb{U}_1{}_0 \ldots {}_0\mathbb{U}_m]\!]\#, w)\\
&\qquad \rightarrow \mathrm{F}(z_1, \ldots, z_n, w).
\end{aligned}
$$

There are many other ways of transforming procedures besides this simple introduction of constants, but it was discovered that the introduction of constants, rather remarkably, is sufficient. That is, in the right context it allows us to establish similar results about other procedure transformation techniques without the necessity for tedious procedure writing. We do so in §6.7.

6.6 INDEXES

We have been using the alphabet L' to "talk about" computations over **str**(L). Now we use coding techniques to compress everything into the single alphabet L, which is a subalphabet of L'. In Chapter 4, §4.5 (if L has more than one letter) and §4.6 (if L has one letter) we defined codings T of L^* in L'^* and U of L'^* in L^*, both of which gave us conservative implementations. T, in fact, was just the *identity map* on L^* (an injection of L^* in L'^*). For the rest of this section, T and U denote these codings, and also the associated conservative implementations. Both T and U assign to each word a single code word, so we talk about *the* code of a word, rather than *a* code, and we write $y = X^U$ rather than $y \in X^U$.

Suppose that **R** is a generated relation of EFS(**str**(L)), say it is the output of procedure R. As usual, R can be thought of as a word of L'^*, and then R^U is a word of L^*. Call R^U an *index* for **R** (we will modify this definition slightly below). We should think of an index for **R** as instructions for generating **R**, but coded into the alphabet L. Since many procedures can have the same output, **R** will have many indexes in L^*.

Now consider the following relation, connecting indexes with members of the relation being indexed. We establish the important result that it is a *generated* relation of EFS(**str**(L)).

UNIV$(z, x_1, \ldots, x_n) \Leftrightarrow z$ is an index for an n-place relation on L^* and $\langle x_1, \ldots, x_n\rangle$ is in that relation $\Leftrightarrow$ $(\exists y)(z = y^U \wedge$ GENERATES_N$(y, x_1, \ldots, x_n))$.

When we showed that U was a conservative implementation in Chapter 4, we did so by showing the UT-coding was generated in EFS(**str**(L')). That is, the relation $z = y^{UT}$ is generated. But y^{UT} is just y^U since T is the identity map on L^*, and so the relation $z = y^U$ is generated in EFS(**str**(L')). Also, by item 27 from §6.4, the following relation is generated: GENERATES_N($y, x_1, \dots, x_n$)$\Leftrightarrow y$ is an EFS(**str**(L)) procedure that outputs n-tuples, and $\langle x_1, \dots, x_n \rangle$ is in the output of y. It follows from the closure properties of generated relations that UNIV($z, x_1, \dots, x_n$) is generated in EFS(**str**(L')). Next we move things to EFS(**str**(L)).

Suppose we think of UNIV($z, x_1, \dots, x_n$) as a relation on L^*. Since the T-coding simply codes members of L^* by themselves, $\text{UNIV}^T = \text{UNIV}$. But we showed in Chapter 4 that T was a *conservative* implementation, which means that UNIV is generated in EFS(**str**(L)) if and only if UNIV^T is generated in EFS(**str**(L')). It follows that UNIV must actually be a generated relation of EFS(**str**(L)).

UNIV is called a *universal* relation because we can get every n-place generated relation from it. Let **R** be any generated n-place relation on L^*. Then **R** has an index i, and $\langle x_1, \dots, x_n \rangle \in \mathbf{R} \Leftrightarrow$ UNIV($i, x_1, \dots, x_n$). That is, every generated n-place relation on L^* is a *section* of UNIV. Further, UNIV itself is a generated relation of EFS(**str**(L)) so the n-place relation $S(x_1, \dots, x_n) \Leftrightarrow$ UNIV($i, x_1, \dots, x_n$) is *always* generated whether i is the U-code for a procedure or not. This suggests we broaden the definition of index. From now on, the following is the official version.

Definition. Every word of L^* is an *index* for a generated n-place relation. Specifically, i is an index for the relation $\{\langle x_1, \dots, x_n \rangle \mid \text{UNIV}(i, x_1, \dots, x_n)\}$, where UNIV is the relation shown to be generated in EFS(**str**(L)) above. We write R_i for the n-place relation with index i. Thus we treat the following as equivalent. $\langle x_1, \dots, x_n \rangle \in R_i \Leftrightarrow R_i(x_1, \dots, x_n) \Leftrightarrow \text{UNIV}(i, x_1, \dots, x_n)$.

The following summarizes the work so far.

Indexing Theorem 6.6.1. *Every generated n-place relation of EFS*(**str**(L)) *has an index. More precisely, there is an* n + 1*-place relation UNIV on* L* *such that:*

1. *UNIV is a generated relation of EFS*(**str**(L)), *and*
2. *if **R** is any generated* n*-place relation of EFS*(**str**(L)), *then for some* $\mathrm{i} \in \mathrm{L}^*$, ***R***$(\mathrm{x}_1, \dots, \mathrm{x}_n) \Leftrightarrow$ *UNIV*$(\mathrm{i}, \mathrm{x}_1, \dots, \mathrm{x}_n)$.

We have been working with n-place relations. Of course, n is arbitrary. We will use the same notation, UNIV, R_i, etc. no matter what value of n is chosen. You can tell what n we have in mind by seeing how many places the relations are shown as having. Finally, the work from §6.5 on transforming procedures by introducing constants gives us a similar transformation on indexes. The result is a version of what has come to be known as the $s-m-n$ theorem (substituting m constants, leaving n variables).

s-m-n Theorem 6.6.2. *Let **R** be a generated* n + m*-place relation of EFS*(**str**(L)). *There is an* n*-argument computable function* f *such that* $f(c_1, \ldots, c_n)$ *is an index for the relation* $\{\langle x_1, \ldots, x_m\rangle \mid \boldsymbol{R}(c_1, \ldots, c_n, x_1, \ldots, x_m)\}$. *That is,* $R_{f(c_1,\ldots,c_n)}(x_1, \ldots, x_m) \Leftrightarrow \boldsymbol{R}(c_1, \ldots, c_n, x_1, \ldots, x_m)$.

Exercise 6.6.1. Derive this theorem from Proposition 6.5.1.

Exercise 6.6.2. Show there is a number N such that every computable function of one argument can be computed by some register machine program using not more than N registers.

6.7 SOME POSITIVE RESULTS

Before we turn to things that can not be computed we give some simple consequences of the work so far, concerning things that can be.

At the end of §6.5 we said that other methods of transforming procedures, besides introducing constants, did not need special treatment; the ability to deal with them would follow easily. Now we justify that claim. Suppose, as an example, that we define a relation T from R and S by $T(x, y) \Leftrightarrow R(x, y) \wedge S(x, y)$. We showed in Chapter 2 that if R and S were generated, so was T. Now we show how to calculate an index for T from indexes for R and S.

Fact. There is a computable function f such that $R_{f(i,j)}(x, y) \Leftrightarrow R_i(x, y) \wedge R_j(x, y)$. In words, if we have indexes i and j for two generated relations, then $f(i, j)$ gives us an index for their conjunction.

Proof of fact: Use the (3-place) universal relation from §6.6, and define a relation S by $S(z, w, x, y) \Leftrightarrow \text{UNIV}(z, x, y) \wedge \text{UNIV}(w, x, y)$. Since UNIV is generated, and generated relations are closed under $\wedge$, S is generated. Now apply the *s-m-n* theorem (6.6.2). There is a computable function f such that $R_{f(z,w)}(x, y) \Leftrightarrow S(z, w, x, y)$. That is, $R_{f(z,w)}(x, y) \Leftrightarrow \text{UNIV}(z, x, y) \wedge \text{UNIV}(w, x, y) \Leftrightarrow R_z(x, y) \wedge R_w(x, y)$.

Exercise 6.7.1. Show there is a computable function f such that

$$R_{f(i,j)}(x, y) \Leftrightarrow (\forall z \text{ SUBWORD } y) R_i(x, z) \wedge R_j(x, y).$$

Exercise 6.7.2. We really have a whole family of indexings, one for 1-place relations, one for 2-place relations, and so on. A 1-place relation (say) can be turned into a 2-place relation by adding an extra position and allowing its occupant to be anything. Show that there is a corresponding function on indexes. That is, show there is a computable function f such that $R_x(y) \Leftrightarrow R_{f(x)}(y, z)$.

We got our indexes by coding EFS procedures. It is also possible to get an indexing by coding IMP or register machine programs for semicharacteristic functions, or by coding LOG definitions. Likewise, we could have used a code based on whatever your favorite language happens to be. Thus, one can develop a whole array of apparently unrelated indexings for the generated relations. One expects to be able to move algorithms between programming languages. Further, one expects that this can be done in a "mechanical" way. This comes down to saying that if we have two indexings, arising by coding different programming languages, there should be computable functions for turning an index of one kind into an equivalent index of the other, and conversely. In order to establish a result of this kind in some generality one must give some criteria for being an "acceptable" indexing. The criteria that people have discovered to work is: we want the Indexing and *s-m-n* theorems to hold for it.

Definition. Suppose we have a family of relations, one 2-place, one 3-place, . . . , all generated, and all denoted **U** for convenience. We say that i is a *U-index* for the n-place relation **R** if $\mathbf{R}(x_1, \ldots, x_n) \Leftrightarrow \mathbf{U}(i, x_1, \ldots, x_n)$. We say that *U* gives an *acceptable* indexing if:

1. every generated relation has a **U**-index, and
2. if **R** is any generated $n+m$-place relation, then there is a computable function f such that $f(c_1, \ldots, c_n)$ is a **U**-index for the relation $\{\langle x_1, \ldots, x_m\rangle \mid \mathbf{R}(c_1, \ldots, c_n, x_1, \ldots, x_m)\}$.

What we showed in §6.6 is that the UNIV-indexing is acceptable (hence an acceptable indexing exists). Well, if we had used coded IMP programs as indexes instead, it is still possible to show an acceptable indexing results. This is the case for any other reasonable choice of programming language. Indeed, providing us with an acceptable indexing is a good definition of being a reasonable programming language. The work earlier in this section used nothing more than that the UNIV-indexing is acceptable, and the same is true of the remainder of the chapter. The result below essentially says any two acceptable indexings are intertranslatable, so any one will serve as well as any other.

Proposition 6.7.1. *Suppose we have two acceptable indexings, a* U*-indexing and a* V*-indexing. Let us write* R_i *for the* (n*-place*) *relation with* U*-index* i. *That is,* $R_i(x_1, \ldots, x_n) \Leftrightarrow U(i, x_1, \ldots, x_n)$. *Similarly, we write* S_i *for the relation with* V*-index* i: $S_i(x_1, \ldots, x_n) \Leftrightarrow V(i, x_1, \ldots, x_n)$. *Then, there are computable functions* f *and* g *such that* $R_i = S_{f(i)}$ *and* $S_i = R_{g(i)}$.

Proof. Since V is an indexing, the relation $V(z, x_1, \ldots, x_n)$ is generated. By the *s-m-n* property applied to the U-indexing, there is a computable function g such that $R_{g(z)}(x_1, \ldots, x_n) \Leftrightarrow V(z, x_1, \ldots, x_n)$. But this says that $R_{g(z)}(x_1, \ldots, x_n) \Leftrightarrow S_z(x_1, \ldots, x_n)$. The other direction is similar.

6.8 COMPLEMENTS

Throughout this book we have frequently treated relations in complementary pairs. EQUAL and NOT_EQUAL on words, for instance, or MEMBER_OF and NOT_MEMBER_OF on sets. We are about to establish why we had to be so careful about making sure important generated relations had generated complements: it doesn't always happen. This is the first of a series of results about what can *not* be done with computer languages.

Definition. If **R** is an n-place relation on a domain **D**, by the *complement* of **R** (relative to **D**) we mean the relation $\mathbf{R}^c = \{\langle x_1, \ldots, x_n\rangle \mid x_1, \ldots, x_n \in \mathbf{D}$ and $\langle x_1, \ldots, x_n\rangle \notin \mathbf{R}\}$. The complement of **R** is the n-place relation on **D** that holds precisely when **R** does not.

Theorem 6.8.1. *There is a set of words that is generated in EFS*(**str**(L)) *but whose complement is not generated.*

Proof. Let UNIV be the universal relation (for 1-place relations) from Theorem 6.6.1. Define S by $S(x) \Leftrightarrow \text{UNIV}(x, x) \Leftrightarrow R_x(x)$. Since UNIV is generated, so is S. Suppose the complement, S^c, were generated. Then it would have an index, so for some i we would have $S^c(x) \Leftrightarrow R_i(x)$. Combining things, $R_i(x) \Leftrightarrow S^c(x) \Leftrightarrow \text{not } S(x) \Leftrightarrow$ not $R_x(x)$. This is to hold for every value of x; take x to be i. Then $R_i(i) \Leftrightarrow$ not $R_i(i)$. This is impossible, so the complement of S could not have been generated.

Exercise 6.8.1. Show that for each n there is an n-place relation on L^* that is generated but does not have a generated complement.

The theorem just established was for EFS(**str**(L)), but similar results follow easily for many other data structures including **integer, tree**(**A**); and **set**(**A**) under reasonable assumptions about **A**.

Corollary 6.8.2. *Suppose that* $\mathbf{S} = \langle \mathbf{D}; \mathbf{R}_1, \ldots, \mathbf{R}_n\rangle$ *is a data structure, and* T *is a conservative implementation of* **str**(L) *in* **S**. *Then there is a subset of* **D** *that is generated in EFS*(**S**) *but whose complement is not.*

Proof. Since T is an implementation, by definition, EQUAL^T is generated in EFS(**S**), where EQUAL represents the equality relation on L^*. Define $\text{CODE}_L(x) \Leftrightarrow \text{EQUAL}^T(x, x)$. Then CODE_L is also generated, and clearly it represents the set of T-codes for members of L^*.

Now suppose that in EFS(**S**), every generated set has a generated complement. We show that this leads to a contradiction. Let S be as in the proof of Theorem 6.8.1. Then S represents a subset of L^* that is generated but whose complement is not. Consider the following three subsets of **D**: $(S)^T$, $(S^c)^T$, $(\text{CODE}_L)^c$. (See Figure 6.1.)

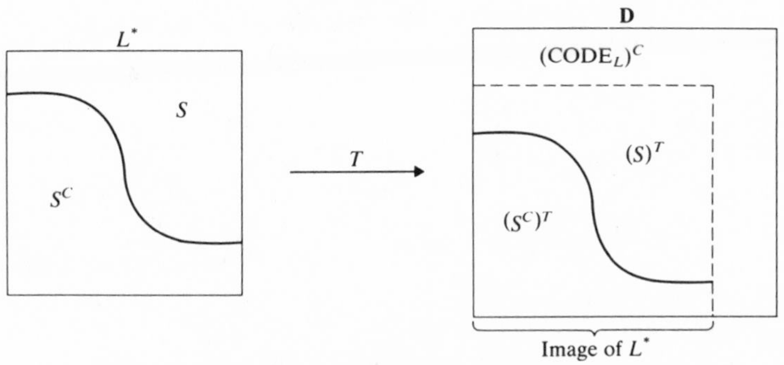

Figure 6.1

(*Note:* S^c means the complement in L^*, while $(\mathrm{CODE}_L)^c$ means the complement in **D**.) Since every member of L^* is in exactly one of S or S^c, every T-code is in exactly one of $(S)^T$ or $(S^c)^T$. And, of course, every non-T-code is in $(\mathrm{CODE}_L)^c$. It follows that the three sets, $(S)^T$, $(S^c)^T$, and $(\mathrm{CODE}_L)^c$, do not overlap, and together include all members of **D**.

S is generated in EFS(**str**(L)), so by Theorem 4.3.1, $(S)^T$ is generated in EFS(**S**). We have already seen that CODE_L is generated in EFS(**S**), and we are supposing closure under complementation, hence $(\mathrm{CODE}_L)^c$ must be generated. Then, since generated relations are closed under $\vee$, $(S)^T \cup (\mathrm{CODE}_L)^c$ must be generated. But this is the complement of $(S^c)^T$, so $(S^c)^T$ too must be generated. Then, since T is a conservative implementation, S^c must be generated in EFS(**str**(L)), and we know it is not. Hence the generated relations of EFS(**S**) cannot be closed under complement.

The origins of the proof of Theorem 6.8.1 can be traced back to Cantor's argument that there can be no one-one, or even many-one mapping from a set to its power set, the collection of all its subsets. The following is intended to bring out the analogy.

Exercise 6.8.2. Let **A** be a set, and suppose there is a many-one correspondence between members of **A** and subsets of **A**. That is, suppose there is a function f that assigns to each member x of **A** some subset $f(x)$ of **A**, such that every subset of **A** is in the range of f. Define a relation **U** on **A** by $\mathbf{U}(x, y) \Leftrightarrow y \in f(x)$. Then $f(x) = \{y \mid \mathbf{U}(x, y)\}$. Call x an *index* for the set $f(x)$, and write $R_x(y)$ for $\mathbf{U}(x, y)$. Our hypothesis amounts to this: *every* subset of **A** has an index. Now derive a contradiction.

6.9 THE HALTING PROBLEM

Definition. A relation **R** (on a domain **D**) is called *decidable* or is said to *have a decision procedure* if its characteristic function is computable. Recall that the characteristic function $c_{\mathbf{R}}$ has value $\bar{1}$ on members of **R** and $\bar{0}$ on members of **D** not in **R**.

A decidable relation is one for which there is some computational method of figuring out what belongs and what doesn't. When we say a characteristic function is computable we can mean either that its graph is generated by an EFS procedure or that there is an IMP program for it. By Theorem 5.12.3 these are equivalent. By Proposition 5.12.4 a decidable relation is also one such that both it and its complement are generated by EFS procedures, so in §6.8 we gave an example of a relation that is not decidable.

The *halting problem* for IMP(L), or equivalently for register machines over L, is the problem of deciding which programs will halt on which inputs. More formally, consider the relation $\{\langle x, y\rangle \mid x$ is an IMP(L) program that halts on input $y\}$. Is this a decidable relation? To a large extent, computability theory began with Turing's 1936 proof that the halting problem was unsolvable: The relation in question does not have a decision procedure. We do not prove this result in the form we stated because it talks about inputs (words over L) and IMP programs (words over a larger alphabet L'). This means we must either do everything using L', to have IMP programs, or else code IMP programs into the alphabet L. Either way involves substantial effort that can be avoided. Suppose that there were some single IMP program for which we could not decide on what inputs it halted. If there were one such program, then certainly the general halting problem would not be decidable either.

Theorem 6.9.1. *There is an IMP*(L) *program PROG such that*

$\{\mathrm{x} \in \mathrm{L}^* \mid$ *PROG halts when given* x *as input*$\}$

is not decidable.

Proof. By Theorem 6.8.1 there is a set S of words over L such that S is generated but its complement, S^c, is not. Since S is generated, by Proposition 5.12.1 there is an IMP program, say PROG, for the semicharacteristic function of S (as the proposition is stated, PROG is an IMP(L') program, but we noted at the end of Chapter 5, §5.12 that the alphabet L' could be replaced by L). Then PROG halts when members of S are used as input and does not halt when given members of S^c. That is, $\{x \in L^* \mid \text{PROG halts when given } x \text{ as input}\} = S$, and this is not a decidable set.

Exercise 6.9.1. Show there are two subsets A and B of L^* that cannot be computationally separated in the sense that for no decidable set S do we have $A \subseteq S$ and $B \subseteq S^c$. *Hint*: Let $A = \{i \mid R_i(i, \bar{0})\}$ and $B = \{i \mid R_i(i, \bar{1})\}$. Suppose a "separating" set S existed for this pair A and B. (See Figure 6.2.) The characteristic function c_s of S would have a generated graph, hence this graph would have an index, say i_0. Question: Where is i_0?

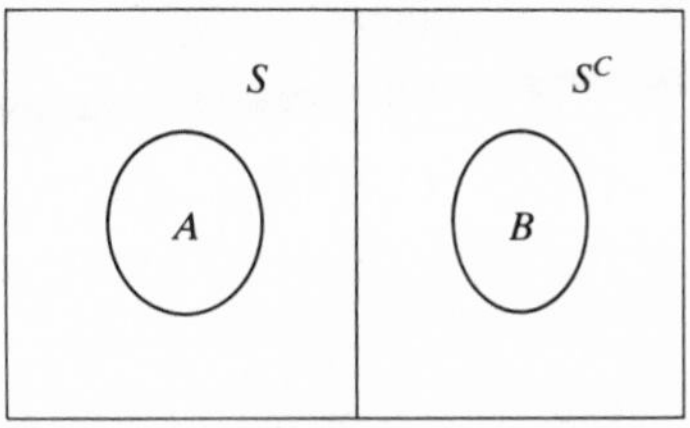

Figure 6.2

6.10 THE UNIVERSAL QUANTIFIER

In Chapter 5 we introduced a language LOG(L) in which one describes relations using mechanisms of formal logic. We showed that EFS, LOG, IMP, and register machine languages were all equivalent. The universal quantifier, $\forall$, was not made part of the machinery of LOG(L) for intuitively obvious reasons: To verify $(\forall x)R(x, c)$ requires that infinitely many items be checked, $R(x, c)$ for each x, and this is contrary to the finitistic nature of computers. But we might wonder if the universal quantifier can be synthesized indirectly. Given the machinery of LOG(L) is there some construction that will act like a universal quantifier? In this section we show that it is not the case.

Definition. Suppose that $R(x_1, \ldots, x_n)$ is a relation. When dealing with LOG(L) we write $\neg R(x_1, \ldots, x_n)$ to denote the complementary relation, R^c in earlier sections. That is, $\neg R(x_1, \ldots, x_n) = \{\langle x_1, \ldots, x_n \rangle \mid R(x_1, \ldots, x_n)$ does *not* hold$\}$.

The symbol $\neg$ is called *negation,* and is standard logical notation. It is not part of the machinery of LOG(L). In certain cases, if $R(x_1, \ldots, x_n)$ has a LOG(L) definition, so will $\neg R(x_1, \ldots, x_n)$, though Theorem 6.8.1 tells us it doesn't always happen. There are some general principles governing the behavior of negation, which we state now.

1. On the atomic level, the relation $\neg$CON(x, y, z) has a LOG(L) definition. The following will do.

   $$\neg\text{CON}(x, y, z) \Leftrightarrow (\exists w)(\text{CON}(x, y, w) \wedge \text{NOT\_EQUAL}(w, z)).$$

   Better yet, it actually has a LOG$^-$(L) definition (see Chapter 5, §5.9).

   $$\neg\text{CON}(x, y, z) \Leftrightarrow \text{NOT\_BEGINS}(x, z) \vee [\text{BEGINS}(x, z) \wedge (\exists w \text{ SUBWORD } z)(\text{CON}(x, w, z) \wedge \text{NOT\_EQUAL}(y, w))]$$

2. Suppose that there are LOG(L) definitions for the relations $\neg P(x_1, \ldots, x_n)$ and $\neg Q(x_1, \ldots, x_n)$. Then there are LOG(L) definitions for

   $$\neg[P(x_1, \ldots, x_n) \wedge Q(x_1, \ldots, x_n)]$$
   $$\neg[P(x_1, \ldots, x_n) \vee Q(x_1, \ldots, x_n)].$$

Exercise 6.10.1. Show this by showing

$$\neg[P(x_1, \ldots, x_n) \wedge Q(x_1, \ldots, x_n)] \text{ is } \neg P(x_1, \ldots, x_n) \vee \neg Q(x_1, \ldots, x_n),$$
$$\neg[P(x_1, \ldots, x_n) \vee Q(x_1, \ldots, x_n)] \text{ is } \neg P(x_1, \ldots, x_n) \wedge \neg Q(x_1, \ldots, x_n).$$

3. Suppose that there is a LOG(L) definition for $\neg P(y, x_1, \ldots, x_n)$. Then there are LOG($L$) definitions for

$$\neg(\exists y \text{ SUBWORD } w)P(y, x_1, \ldots, x_n)$$
$$\neg(\forall y \text{ SUBWORD } w)P(y, x_1, \ldots, x_n).$$

Exercise 6.10.2. Show this by showing

$$\neg(\exists y \text{ SUBWORD } w)P(y, x_1, \ldots, x_n) \text{ is } (\forall y \text{ SUBWORD } w)$$
$$\neg P(y, x_1, \ldots, x_n),$$
$$\neg(\forall y \text{ SUBWORD } w)P(y, x_1, \ldots, x_n) \text{ is } (\exists y \text{ SUBWORD } w)$$
$$\neg P(y, x_1, \ldots, x_n).$$

What we have shown so far is that as we construct a LOG(L) definition from CON that only uses $\wedge$, $\vee$, and bounded quantification we can, in parallel, construct a LOG(L) definition of the complementary relation, again just using $\wedge$, $\vee$, and bounded quantification. We have shown the following.

Proposition 6.10.1. *If a relation has a LOG^-(L) definition, its complement has a LOG^-(L) definition. That is, LOG^-(L) is closed under* $\neg$.

The universal quantifier is not part of the given machinery of LOG(L), but we can still characterize its behavior on relations. If $P(y, x_1, \ldots, x_n)$ denotes a relation, $(\forall y)P(y, x_1, \ldots, x_n) = \{\langle x_1, \ldots, x_n\rangle \mid P(y, x_1, \ldots, x_n)$ holds for *every* word $y\}$.

Exercise 6.10.3. Suppose that $P(y, x_1, \ldots, x_n)$ is a relation. Show

$$\neg(\exists y)P(y, x_1, \ldots, x_n) \text{ is } (\forall y)\neg P(y, x_1, \ldots, x_n),$$
$$\neg(\forall y)P(y, x_1, \ldots, x_n) \text{ is } (\exists y)\neg P(y, x_1, \ldots, x_n).$$

By this exercise and what was said earlier, if the universal quantifier were indirectly available in LOG(L) then LOG(L) would be closed under complementation. But, by the work in Chapter 5, §5.10, the LOG definable relations are the EFS generable ones, and by Theorem 6.8.1, the generated relations are not closed under complementation. Consequently, the universal quantifier cannot be simulated in LOG(L).

Further, since there is a LOG(L) definable relation whose complement is not definable, Proposition 6.10.1 gives us the following.

Proposition 6.10.2. *The LOG^-(L) definable relations constitute a proper subclass of the LOG(L) definable relations.*

6.11 THE SECOND RECURSION THEOREM

Suppose that c is one of the letters of the alphabet L, and we want to show the set consisting of words made up of just c's is generated in EFS(**str**(L)). Of course, in this simple case we could easily write a procedure to do the job but we want to illustrate other ways of thinking about the problem, ways that are useful in more complicated cases. Call the set we are after S. Thus, $S = \{\Delta, c, cc, ccc, \ldots\}$. S can be characterized by saying that it is the unique set such that

$$S(x) \Leftrightarrow \text{EQUAL}(x, \Delta) \vee (\exists y)[S(y) \wedge \text{CON}(y, c, x)].$$

We temporarily call this the *S-condition*. Thus, the problem is to show the set that meets the S-condition must be a generated one.

Exercise 6.11.1. Verify the assertion that S is the unique set meeting the S-condition:

(a) Show that S meets the S-condition.
(b) Show by induction on length that every word made up entirely of c's is in any set meeting the S-condition.
(c) Show by induction on length that every word in a set meeting the S-condition must consist entirely of c's.

The S-condition is a recursive one, S appears on both sides. S is being characterized in terms of itself. One way to "open up" this recursive condition is to think of the two occurrences of S as really different. Say we use different letters, A and B for them. Consider the revised condition

$$B(x) \Leftrightarrow \text{EQUAL}(x, \Delta) \vee (\exists y)[A(y) \wedge \text{CON}(y, c, x)].$$

If we have some set A that is a candidate for S, and we use this to define B, B should be a better candidate. For example, if $A = \{\Delta, c, cc\}$ then this definition will make $B = \{\Delta, c, cc, ccc\}$. What we are after is the "best" candidate, the set S that can serve as both the A and B in this definition. We have replaced the original recursive characterization of S by a definition of an operator.

$$B = \Phi(A) = \{x \mid \text{EQUAL}(x, \Delta) \vee (\exists y)[A(y) \wedge \text{CON}(y, c, x)]\}.$$

The set S we want is one such that $\Phi(S) = S$, a fixed point of Φ. In general, operators like this have many fixed points, and it will be the smallest among them that we want, because it has nothing "extra" in it. In this case however, Φ has only a single fixed point by Exercise 6.11.1.

The First Recursion Theorem (3.10.1) tells us that if we can write an input accepting procedure for Φ, then the set S will be a generated one. Such a procedure is easily written. Using the methods for turning LOG(L) definitions into EFS(**str**(L)) procedures we get

```
B(1) INPUT A (1):
    B(Δ);
    A(y)→CON(y, c, x)→B(x).
```

Now the appropriateness of the name *Recursion Theorem* should be a little clearer. It says that by finding the least fixed point of this operator we are finding a set satisfying our original recursive S-condition.

There is another, quite different, way of showing that generated sets satisfying recursive characterizations must exist. It is also called a recursion theorem and it, like the First Recursion Theorem, is due to Kleene. We state and prove it, then we show how it applies to our recursive condition for S above.

Second Recursion Theorem 6.11.1. *Let* t *be any function from* L* *to* L* *that is total and computable. That is,* (1) t *is defined on every word in* L*, *and* (2) *the graph of* t *is generated. Then there is some index* i *such that* $\mathrm{R}_{\mathrm{t(i)}}(\mathrm{x}_1, \ldots, \mathrm{x}_n) \Leftrightarrow \mathrm{R}_{\mathrm{i}}(\mathrm{x}_1, \ldots, \mathrm{x}_n)$.

In words, for some i, both i and $t(i)$ are indexes of the same generated n-place relation. Thus we have another "fixed point" theorem: moving from i to $t(i)$ may change indexes, but it leaves fixed the relation being indexed.

Proof. Define an $n+1$-place relation S by $S(u, x_1, \ldots, x_n) \Leftrightarrow \mathrm{UNIV}(u, x_1, \ldots, x_n, u) \Leftrightarrow R_u(x_1, \ldots, x_n, u)$. Since UNIV is generated, so is S. Then by the *s-m-n* theorem (6.6.2) there is a (total) function g with generated graph such that

$$R_{g(u)}(x_1, \ldots, x_n) \Leftrightarrow S(u, x_1, \ldots, x_n) \Leftrightarrow R_u(x_1, \ldots, x_n, u) \qquad (*)$$

Since the graphs of both t and g are generated so is the graph of the (total) composite function $h(y) = t(g(y))$ (Exercise 1.5.4). Say the graph of h is represented by H. Then the following relation is a generated one: $W(x_1, \ldots, x_n, y) \Leftrightarrow (\exists s)[H(y, s) \wedge \mathrm{UNIV}(s, x_1, \ldots, x_n)] \Leftrightarrow \mathrm{UNIV}(t(g(y)), x_1, \ldots, x_n) \Leftrightarrow R_{t(g(y))}(x_1, \ldots, x_n)$. Let w be an index for W. Then, by definition,

$$R_w(x_1, \ldots, x_n, y) \Leftrightarrow W(x_1, \ldots, x_n, y) \Leftrightarrow R_{t(g(y))}(x_1, \ldots, x_n) \quad (**)$$

Finally, let $i = g(w)$. We claim that i is the index we want. The verification is as follows.

$$\begin{aligned} R_i(x_1, \ldots, x_n) &\Leftrightarrow R_{g(w)}(x_1, \ldots, x_n) && \text{(def of } i\text{)} \\ &\Leftrightarrow R_w(x_1, \ldots, x_n, w) && \text{(by *)} \\ &\Leftrightarrow R_{t(g(w))}(x_1, \ldots, x_n) && \text{(by **)} \\ &\Leftrightarrow R_{t(i)}(x_1, \ldots, x_n) && \text{(def of } i\text{).} \end{aligned}$$

Now we return to the example with which we began this section. We want to find a generated set S meeting the S-condition:

$$S(x) \Leftrightarrow \mathrm{EQUAL}(x, \Delta) \vee (\exists y)[S(y) \wedge \mathrm{CON}(y, c, x)].$$

In using the *First* Recursion Theorem we replaced the S on the right by *any* set A and thought of the result as a definition of a set B. This time we replace the S on the right by a *generated* set. Since every generated set has an index, we can write this as the following: a definition of B given R_z.

$$B(x) \Leftrightarrow \text{EQUAL}(x, \Delta) \vee (\exists y)[R_z(y) \wedge \text{CON}(y, c, x)].$$

Since the definition only uses machinery from LOG(L), it follows that B must be another generated set. Then B also has indexes, and we can calculate one from the index z for R_z as follows. Let

$$\begin{aligned} C(z, x) &\Leftrightarrow \text{EQUAL}(x, \Delta) \vee (\exists y)[R_z(y) \wedge \text{CON}(y, c, x)] \\ &\Leftrightarrow \text{EQUAL}(x, \Delta) \vee (\exists y)[\text{UNIV}(z, y) \wedge \text{CON}(y, c, x)]. \end{aligned}$$

C is a generated relation; by the *s-m-n* theorem there is a function t whose graph is generated, such that $R_{t(z)}(x) \Leftrightarrow C(z, x)$. That is, $t(z)$ is an index for the set we were calling B above, and

$$R_{t(z)}(x) \Leftrightarrow \text{EQUAL}(x, \Delta) \vee (\exists y)[R_z(y) \wedge \text{CON}(y, c, x)].$$

Now, the *Second* Recursion Theorem says there is some index i such that $R_{t(i)}(x) \Leftrightarrow R_i(x)$. The claim is that R_i is the set $S = \{\Delta, c, cc, ccc, \ldots\}$. But this is easy to see.

$$\begin{aligned} R_i(x) &\Leftrightarrow R_{t(i)}(x) \\ &\Leftrightarrow \text{EQUAL}(x, \Delta) \vee (\exists y)[R_i(y) \wedge \text{CON}(y, c, x)]. \end{aligned}$$

And Exercise 6.11.1 tells us that R_i must be S since the recursive S-condition only has one solution. We have verified that S is generated, by showing it has an index i.

Thus, both recursion theorems can be used to show recursive characterizations of relations have generated solutions. Apart from this common application, the two have very different flavors in general. Indeed, the consequences of the Second Recursion Theorem can be quite remarkable, as you will soon see.

Exercise 6.11.2. In Exercise 3.12.5 you were to use the First Recursion Theorem to show the graph of a certain function was generated. Redo that exercise, but using the Second Recursion Theorem instead.

As a simple but remarkable example of an application of the Second Recursion Theorem we show there is a self-reproducing procedure. More precisely, we show there is a procedure of EFS(**str**(L)) whose only output is an index for that very procedure. Equivalently, we show there is an index i for a generated set such that the only thing in that set is i itself. We show there is a solution to $R_i = \{i\}$.

Consider the relation EQUAL(x, y) on L^* (which we know is generated). By the *s-m-n* theorem there is a function t with generated graph such that $R_{t(x)}(y) \Leftrightarrow \text{EQUAL}(x, y)$. By the Second Recursion Theorem, for some i, $R_{t(i)}(y) \Leftrightarrow R_i(y)$. But then, $R_i(y) \Leftrightarrow R_{t(i)}(y) \Leftrightarrow \text{EQUAL}(i, y)$. That is, the only y such that $R_i(y)$ must be (equal to) i.

Exercise 6.11.3. Suppose we use the *s-m-n* theorem, as above, to get a function f with generated graph such that $R_{f(x)}(y) \Leftrightarrow \text{EQUAL}(x, y)$. Then we use the *s-m-n* theorem again to get a function g with generated graph such that $R_{g(z)}(w) \Leftrightarrow \text{EQUAL}(f(z), w) \Leftrightarrow \langle z, w\rangle$ is in the graph of f. By the Second Recursion Theorem, for some i, $R_{g(i)}(w) \Leftrightarrow R_i(w)$. Now set $j = f(i)$. Show $R_i = \{j\}$ and $R_j = \{i\}$.

It is not hard to see that, by our proof of the *s-m-n* Theorem, the function f in the exercise above will have the property that $f(x) \neq x$ for all x (indeed, $f(x)$ will always be a longer word than x). Then i and j must be different in the exercise. So, in effect, we have created two distinct procedures (indexed by i and j) each of which outputs an index for the other.

Exercise 6.11.4. In §6.7 we defined the notion of an acceptable indexing. Now show that, given any two acceptable indexings, some i must index the same relation in both. *Hint:* See Proposition 6.7.1.

The Second Recursion Theorem can be strengthened. It says that, given a function t with generated graph, an index i exists such that $R_{t(i)} = R_i$. In fact, it is possible to *find* such an index i, given t. But how can t be given to us? It has a generated graph, which thus must have an index, j. We can calculate i from j.

Theorem 6.11.2. *There exists a function* k *with generated graph whose properties are as follows. If* t *is any generated total function, if* j *is any index for the graph of* t, *and if* $\mathrm{i} = \mathrm{k}(\mathrm{j})$, *then* $\mathrm{R}_{\mathrm{t(i)}}(\mathrm{x}_1, \ldots, \mathrm{x}_n) \Leftrightarrow \mathrm{R}_{\mathrm{i}}(\mathrm{x}_1, \ldots, \mathrm{x}_n)$.

Proof. The proof of Theorem 6.11.1 can be modified to produce this stronger result. Let g be the function produced in that proof, let G represent its graph, and let T represent the graph of the function t. The relation W constructed in the proof of Theorem 6.11.1 has the following equivalent characterization:

$$W(x_1, \ldots, x_n, y) \Leftrightarrow (\exists r)(\exists s)[G(y, r) \wedge T(r, s) \wedge \text{UNIV}(s, x_1, \ldots, x_n)]$$

We want to bring indexes for T in explicitly. If the graph of t has index j, then $T(r, s) \Leftrightarrow R_j(r, s) \Leftrightarrow \text{UNIV}(j, r, s)$. So we define W' by

$$W'(j, x_1, \ldots, x_n, y) \Leftrightarrow (\exists r)(\exists s)[G(y, r) \wedge \text{UNIV}(j, r, s) \wedge \text{UNIV}(s, x_1, \ldots, x_n)].$$

W' is generated so, by the *s-m-n* theorem, there is a function f with generated graph such that $R_{f(j)}(x_1, \ldots, x_n, y) \Leftrightarrow W'(j, x_1, \ldots, x_n, y)$. We simply use this $f(j)$ where, in the proof of Theorem 6.11.1, we used the index w for W. Thus, the index i we want is $g(f(j))$ and the function k should be defined as $k(x) = g(f(x))$.

Exercise 6.11.5. Finish the proof by showing, for the function k just defined, if j is an index for the graph of t, then

$$R_{k(j)}(x_1, \ldots, x_n) \Leftrightarrow R_{t(k(j))}(x_1, \ldots, x_n).$$

6.12 RICE'S THEOREM

We have shown there are generated relations without generated complements. It would not be so much of a problem if we could tell which relations these were. Is the set of indexes for relations whose complements are also generated a *decidable* set? Rice's Theorem provides an answer to this question, and to many others of a similar nature.

Definition. Let **C** be a set of words of L^*. We call **C** *n-closed* if, whenever it contains one index for a generated n-place relation it contains all of them. That is, **C** is n-closed provided, if $i \in \mathbf{C}$ and $R_i(x_1, \ldots, x_n) \Leftrightarrow R_j(x_1, \ldots, x_n)$, then $j \in \mathbf{C}$.

Rice's Theorem 6.12.1. *There are only two* n*-closed decidable sets, the empty set* $\varnothing$*, and the set of all words* L^*.

Proof. Let **C** be n-closed, decidable, and suppose $\mathbf{C} \neq \varnothing$ and $\mathbf{C} \neq L^*$; we derive a contradiction. Since $\mathbf{C} \neq \varnothing$, **C** has a member. Choose one, say $a \in \mathbf{C}$. Since $\mathbf{C} \neq L^*$, $\mathbf{C}^c$ is not empty. Choose a member, say $b \in \mathbf{C}^c$. Now define a function t on L^* as follows.

$$t(x) = \begin{cases} b & \text{if } x \in \mathbf{C} \\ a & \text{if } x \notin \mathbf{C} \end{cases}$$

Since **C** is decidable, both **C** and $\mathbf{C}^c$ are generated. It follows that the graph of t is generated; here is a LOG(L) definition for it.

$$t(x) = y \Leftrightarrow [\mathbf{C}(x) \wedge \text{EQUAL}(y, b)] \vee [\mathbf{C}^c(x) \wedge \text{EQUAL}(y, a)].$$

By the Second Recursion Theorem there must be an index i such that $R_{t(i)}(x_1, \ldots, x_n) \Leftrightarrow R_i(x_1, \ldots, x_n)$. That is, both i and $t(i)$ are indexes for the same n-place relation. Then $i \in \mathbf{C}$ if and only if $t(i) \in \mathbf{C}$ since **C** is n-closed. But if $i \in \mathbf{C}$, by definition of t, $t(i) \notin \mathbf{C}$. Hence i cannot be in **C**. A similar argument says that i cannot be in $\mathbf{C}^c$ either, since $\mathbf{C}^c$ is also n-closed. We have reached the desired contradiction, and the theorem is established.

Now we can easily answer the question raised at the beginning of the section. Let **C** be the set of all indexes for n-place relations that are generated and that have generated complements. **C** is not empty since we have provided many examples of generated relations with generated complements. But also, by Theorem 6.8.1 (or Exercise 6.8.1), not every index is in **C**, $\mathbf{C} \neq L^*$. Then, by Rice's Theorem, **C** is not decidable. We cannot tell in general, just by looking at an index, whether or not the complementary relation is generated.

As another example, consider the equivalence problem. A generated relation will have many indexes, not just one. An index is a coded version of an EFS procedure, and we can always add duplicate copies of instructions to procedures without changing behavior. That is, we can easily produce formally different procedures (and hence different indexes) for generating the same set. But this is trivial. If a procedure has duplicate instructions we can recognize that fact and remove them. Are things always so simple? Can we always tell, by looking at indexes, whether or not they index the same relation? Again, the answer is no. The easy argument from Rice's Theorem goes as follows. Choose any index i. If we had a way of deciding when *any* two indexes were indexes for the same n-place relation, then certainly $\{x \mid x$ and i index the same n-place relation$\}$ would be a decidable set. But, not every word is in this set since there are generated n-place relations besides R_i. And the set is not empty, since i is a member. Then the set cannot have a decision procedure, by Rice's Theorem.

Exercise 6.12.1. Let N be the set of all indexes for nonempty generated sets.

(a) Show that N is not decidable.
(b) Give a procedure whose output is N, thus showing that N^c is not generated.

Exercise 6.12.2. Let F be the set of all indexes for generated 2-place relations that are graphs of functions.

(a) Show that F is not decidable.
(b) Show that one of F or F^c, its complement, is generated, thus showing that the other is not generated.

These exercises should not lead you to believe that, in all cases, if a set S of indexes is undecidable, still one of S or S^c will be generated. If S is the set of all indexes for generated sets that have generated complements, our first example in this section, then *neither* S nor S^c are generated. Or again, say S is the set of all indexes for finite sets. An easy application of Rice's Theorem says that S is not decidable, and again neither S nor S^c will be generated. In order to prove these items we need better tools than Theorem 6.12.1; we need results about generated sets, not just decidable ones.

Theorem 6.12.2. *Let* $\mathbf{C}$ *be an* n-*closed, generated set of indexes. If* $\mathrm{a} \in \mathbf{C}$ *and* $\mathrm{R_a} \subseteq \mathrm{R_b}$, *then* $\mathrm{b} \in \mathbf{C}$.

This theorem is a manifestation of the way computational processes ultimately depend on *positive* information. If we conclude $a \in \mathbf{C}$ it must be because certain items *are* in R_a. But those items will also be in any extension R_b of R_a, so we cannot exclude b from $\mathbf{C}$. The proof is not along these lines, but instead uses the Second Recursion Theorem in a way very similar to the proof of Rice's Theorem.

Proof. Let $\mathbf{C}$ be n-closed, generated, and suppose that $a \in \mathbf{C}$ and $R_a \subseteq R_b$. We show $b \in \mathbf{C}$.

Define a relation S as follows. $S(x_1, \ldots, x_n, y) \Leftrightarrow R_a(x_1, \ldots, x_n) \vee [R_b(x_1, \ldots, x_n) \wedge \mathbf{C}(y)]$. Since R_a, R_b, and $\mathbf{C}$ are generated relations, so is S.

Then by the *s-m-n* theorem there is a function f with generated graph such that $R_{f(y)}(x_1, \ldots, x_n) \Leftrightarrow S(x_1, \ldots, x_n, y)$. By the Second Recursion Theorem, for some index i, $R_{f(i)}(x_1, \ldots, x_n) \Leftrightarrow R_i(x_1, \ldots, x_n)$. It follows that, for all $x_1, \ldots, x_n$, $R_i(x_1, \ldots, x_n) \Leftrightarrow R_a(x_1, \ldots, x_n) \vee [R_b(x_1, \ldots, x_n) \wedge \mathbf{C}(i)]$.

Now, suppose that $i \notin \mathbf{C}$. Then the clause $R_b(x_1, \ldots, x_n) \wedge \mathbf{C}(i)$ is always false, and so $R_i(x_1, \ldots, x_n) \Leftrightarrow R_a(x_1, \ldots, x_n)$. This says that i and a index the same relation. Since $a \in \mathbf{C}$ and $\mathbf{C}$ is n-complete, $i \in \mathbf{C}$, contrary to supposition. This establishes that $i \in \mathbf{C}$. But then the clause $\mathbf{C}(i)$ is always true, and so $R_i(x_1, \ldots, x_n) \Leftrightarrow R_a(x_1, \ldots, x_n) \vee R_b(x_1, \ldots, x_n)$. We also have that $R_a \subseteq R_b$, so this further reduces to $R_i(x_1, \ldots, x_n) \Leftrightarrow R_b(x_1, \ldots, x_n)$. This says that i and b index the same relation. Since $i \in \mathbf{C}$ and $\mathbf{C}$ is n-complete, $b \in \mathbf{C}$, which concludes the proof.

Let S be the set of all indexes for finite sets. Then S is 1-complete by definition. If $a \in S$, then R_a must be finite. But by Theorem 6.12.2, if S were generated then for each $a \in S$ we would also have $b \in S$ whenever $R_a \subseteq R_b$, and there are certainly *infinite* generated sets R_b that extend R_a (the set of all words, for instance). So S is not a generated set.

Exercise 6.12.3. Let S be the set of all indexes for generated sets that have generated complements. Show that S is not generated. *Hint:* First show that S contains indexes for all the finite sets.

Theorem 6.12.3. *Let* $\mathbf{C}$ *be an* n-*closed, generated set of indexes. If* $b \in \mathbf{C}$ *then for some finite* R_a *we have* $R_a \subseteq R_b$ *and* $a \in \mathbf{C}$.

This theorem can be thought of as an instance of the *finitary* nature of computations. In order to establish that $b \in \mathbf{C}$, we can only use a finite amount of information about R_b, hence there is some finite subset of R_b to which our argument will also apply.

Proof. In Chapter 5, we showed how to translate register machine programs into EFS procedures. It is an easy exercise, which we leave to you, to extend that work to prove some stronger results. Let f be a (possibly partial) register machine computable function; equivalently f has a generated graph. Then the following relations are generated.

1. $E(x_1, \ldots, x_n, y) \Leftrightarrow$ the register machine trace for $f(x_1, \ldots, x_n)$ has y symbols (thinking of y as a number representative and counting arrow, newline, etc. as symbols)
2. $M(x_1, \ldots, x_n, y) \Leftrightarrow$ the register machine trace for $f(x_1, \ldots, x_n)$ has more than y symbols (possibly it has infinitely many).

Now suppose that $\mathbf{C}$ is n-closed, generated, and $b \in \mathbf{C}$. We produce a finite $R_a \subseteq R_b$ with $a \in \mathbf{C}$.

Since both **C** and R_b are generated, they both have computable semicharacteristic functions. We write s_c and s_{Rb} for them. Then

$$s_{Rb}(x_1, \ldots, x_n) = \begin{cases} 1 & \text{if } \langle x_1, \ldots, x_n \rangle \in R_b \\ \text{undefined otherwise} \end{cases}$$

and similarly for s_c. Define a relation S as follows.

$$S(x_1, \ldots, x_n, y) \Leftrightarrow (\exists z)[\text{the trace for } s_{Rb}(x_1, \ldots, x_n) \text{ has } z \text{ symbols} \\ \wedge \text{ the trace for } s_{\mathbf{C}}(y) \text{ has more than } z \text{ symbols}].$$

S is a generated relation by items 1 and 2 above. Think of $S(x_1, \ldots, x_n, y)$ as saying that we will discover that $\langle x_1, \ldots, x_n \rangle \in R_b$ before we discover that $y \in \mathbf{C}$ (if we ever discover that $y \in \mathbf{C}$). By the *s-m-n* theorem, there is a function f with generated graph such that $R_{f(y)}(x_1, \ldots, x_n) \Leftrightarrow S(x_1, \ldots, x_n, y)$. And, by the Second Recursion Theorem, for some index a, $R_{f(a)}(x_1, \ldots, x_n) \Leftrightarrow R_a(x_1, \ldots, x_n)$. It follows that, for all $x_1, \ldots, x_n$,

$$R_a(x_1, \ldots, x_n) \Leftrightarrow (\exists z)[\text{the trace for } s_{Rb}(x_1, \ldots, x_n) \text{ has } z \text{ symbols} \\ \wedge \text{ the trace for } s_c(a) \text{ has more than } z \text{ symbols}].$$

If $R_a(x_1, \ldots, x_n)$, then for some z the trace for $s_{Rb}(x_1, \ldots, x_n)$ has z symbols, which certainly implies that $R_b(x_1, \ldots, x_n)$. Hence $R_a \subseteq R_b$.

Next, suppose that $a \notin \mathbf{C}$; we derive a contradiction. Well, if $a \notin \mathbf{C}$ the trace for $s_c(a)$ is infinite, and so the relation [the trace for $s_c(a)$ has more than z symbols] is true for *every* number (representative) z. It follows that $R_a(x_1, \ldots, x_n) \Leftrightarrow (\exists z)$[the trace for $s_{Rb}(x_1, \ldots, x_n)$ has z symbols] $\Leftrightarrow R_b(x_1, \ldots, x_n)$. This says that a and b index the same relation. Since $b \in \mathbf{C}$ and $\mathbf{C}$ is n-closed, $a \in \mathbf{C}$, contradicting our supposition.

We have established that $a \in \mathbf{C}$. Finally, we show that R_a is finite. Since $a \in \mathbf{C}$, the trace for $s_c(a)$ is finite. Say it has k symbols. Then the relation [the trace for $s_c(a)$ has more than z symbols] is true exactly when $z < k$. It follows that

$$R_a(x_1, \ldots, x_n) \Leftrightarrow \\ (\exists z)[\text{the trace for } s_{Rb}(x_1, \ldots, x_n) \text{ has } z \text{ symbols} \wedge z < k].$$

There are only a finite number of traces with fewer than k symbols, so $R_a(x_1, \ldots, x_n)$ holds for only a finite number of $\langle x_1, \ldots, x_n \rangle$. This concludes the proof.

Again, let S be the set of all indexes for finite sets; S is 1-complete by definition, and so is S^c. Certainly S^c is not empty so if it were generated it would follow from Theorem 6.12.3 that some index for a finite set must be in S^c, which is impossible. Consequently, S^c is not generated.

Exercise 6.12.4. Let S be the set of all indexes for generated sets that have generated complements. Show that S^c is not generated.

Exercise 6.12.5. For a fixed index i, we showed earlier that the set

$S(i) = \{x \mid x \text{ and } i \text{ index the same } n\text{-place relation}\}$

is not decidable.

(a) Show that $S(i)$ is not generated.
(b) Give an example of an index i for which $S(i)^c$ is not generated.
(c) Give an example of an index for which $S(i)^c$ is generated.

Hint: For (b) and (c), characterize the index i by saying what the set R_i is to be, and choose extreme examples for R_i.

The list of applications of Rice's Theorem is long, even without the refinements brought by Theorems 6.12.2 and 6.12.3. Essentially Rice's Theorem says that any nontrivial set of indexes that we might be interested in will be undecidable. And the only machinery that went into the proof of Rice's Theorem was that of an acceptable indexing (see §6.7). So it applies to any reasonable computing language we might devise. In particular, we could have used coded IMP programs as indexes, this would also give us an acceptable indexing and Rice's Theorem would apply. An easy consequence is: the set of indexes for IMP programs that halt when given some particular word w_0 as input is not a decidable set. Theorem 6.9.1 said there was an IMP(L) program such that, if we varied the input, we could not tell on which inputs it would halt. The present result says, if we keep the input fixed but vary the IMP program, we can not tell which programs will halt on that input. This is another variation on the unsolvability of the halting problem. There is no sure way of picking out those IMP programs that will cause infinite looping.

6.13 BACKGROUND

The oldest major result in computability theory, the one that virtually started the field off, is Turing's proof of the unsolvability of the halting problem [Turing 1936]. Though his proof looks superficially different from the one given here, it is the same in essence. Turing created a mathematical model of a computation device, now known as a Turing machine. Then he showed there was a "universal" Turing machine, one that could simulate all the others. This is the counterpart of our EFS interpreter written in EFS. Finally, he used the universal Turing machine to show the halting problem was unsolvable by an argument much like the one given here in §6.8 to show not every generated set has a generated complement. Turing's important paper is quite readable, and is strongly recommended

It is interesting to note that in the early stages of the development of the language LISP McCarthy wrote a "universal" LISP function, essentially to show how easily it could be done in LISP. It was quickly realized that this function had practical application; it was hand coded and became the first working LISP interpreter. This is the ancestor of the EVAL function found in current LISP's [McCarthy 1981].

The indexing theorem is essentially due to Turing. The *s–m–n* Theorem and many of its applications are due to Kleene. In particular the Second Recursion Theorem is due to Kleene. The result, and its rather mysterious proof, grew out of earlier work on the lambda calculus. A good presentation by Kleene may be found in [Kleene 1952].

In §6.11 we showed there must be a self-reproducing program in EFS (i.e., a program that outputs an index for itself). [Burger, Brill, and Machi 1980] contains interesting examples of self-reproducing programs written in C and LISP.

Rice's Theorem is from [Rice 1953], though the simple proof given here is not the original one.

REFERENCES

[Burger, Brill, and Machi 1980] Self-reproducing programs, J. Burger, D. Brill, and F. Machi, *BYTE,* Vol. 5, No. 8, pp. 72–74, August, 1982.

[Kleene 1952] *Introduction to Metamathematics,* S. Kleene, D. Van Nostrand Co., Princeton, N. J., 1952.

[McCarthy 1981] History of LISP, J. McCarthy, *History of Programming Languages,* Richard L. Wexelblat, editor, Academic Press, New York, pp. 173–185, 1981.

[Rice 1953] Classes of recursively enumerable sets and their decision problems, G. Rice, *Transactions of the American Mathematical Society,* Vol. 74, pp. 358–366, 1953.

[Turing 1936] On computable numbers, with an application to the entscheidungsproblem, A. Turing, *Proc. London Mathematical Society,* Ser. 2, Vol. 42, pp. 230–265, 1936, corrections ibid., Vol. 43, pp. 544–546, 1937. Reprinted in *The Undecidable,* M. Davis, editor, Raven Press, Hewlett, N. Y., pp. 116–154, 1965.

Appendix
ON INDUCTION

A mathematical tool we use very frequently throughout this book is *induction*. In case you are not at home with proofs that use it, we present a brief discussion here.

We take the *natural numbers* to be $0, 1, 2, \ldots$. The number 1 is often used as a starting point in other works, but 0 is more convenient here. The *principle of induction* is a fundamental feature of the natural number system. It can be formulated quite simply as follows.

Definition. Call a set S of natural numbers *inductive* if:

$0 \in S$ and
for any number k, if $k \in S$ then $k + 1 \in S$.

Principle of Induction. *There is only one inductive set of natural numbers, namely the set containing all natural numbers.*

This principle allows us to establish results about all natural numbers using *proofs by induction*. For example, suppose we have a function f from the set of natural numbers to itself that meets the conditions:

$$f(0) = 3$$
$$f(x + 1) = f(x) + 2.$$

There is exactly one such function. Take that for granted. Now suppose we want to prove that, in fact, $f(x) = 2 \cdot x + 3$. We can do this rigorously as follows. Define a set S of natural numbers by:

$n \in S$ if and only if $f(n) = 2 \cdot n + 3$.

If we show that S is inductive, by the principle of induction every natural number will be in S, which means, for every natural number x, $f(x) = 2 \cdot x + 3$. And we show that S is inductive by a straightforward argument.

Basis step: We show that $0 \in S$. But, in fact, we are told that $f(0) = 3$, and certainly $2 \cdot 0 + 3 = 3$. Then, $f(x) = 2 \cdot x + 3$ if $x = 0$, so $0 \in S$ by definition.

Induction step: Suppose that $k \in S$ (induction hypothesis). We show that $k + 1 \in S$. Well, since $k \in S$ we have $f(k) = 2 \cdot k + 3$. Then, by the conditions given for f,

$$\begin{aligned} f(k+1) &= f(k)+2 \\ &= (2\cdot k+3)+2 \quad \text{(by the induction hypothesis)} \\ &= 2\cdot k+5 \\ &= 2\cdot(k+1)+3 \end{aligned}$$

so $f(x) = 2\cdot x+3$ if $x = k+1$, and so $k+1 \in S$. This ends the induction step, and the proof.

Often induction is presented as a method for showing every natural number has some property, say **P**. That is, if one can show 0 has property **P**, and that $k+1$ has property **P** whenever k has property **P**, then every natural number has property **P**. This is equivalent to the principle of induction as we gave it. Just think of the set S as the set of all natural numbers that have the property **P**. But speaking about properties instead of sets may make it easier to follow a particular proof.

There is another version of induction that we use even more frequently than the one above. For lack of a better name we might call it *complete* induction. It is formulated as follows.

Definition. Call a set T of natural numbers *completely inductive* if, for any number k, $k \in T$ provided all natural numbers smaller than k are in T.

Principle of Complete Induction. *There is only one completely inductive set of natural numbers, the one containing every natural number.*

The intuition behind this is slightly trickier than that behind the principle of induction. We can argue informally as follows. Suppose that T is completely inductive. There are no natural numbers smaller than 0, so every natural number smaller than 0 is in T (show me one that isn't). Consequently, since T is completely inductive, $0 \in T$. But 0 is the only natural number smaller than 1, and $0 \in T$, hence $1 \in T$ since T is completely inductive. But 0 and 1 are the only natural numbers smaller than 2, hence etc. This discussion was meant to be suggestive, not rigorous. That "etc." at the end is the problem. In fact, getting around the use of that word is precisely what we want complete induction for, so our informal discussion is really circular.

The principle of complete induction should not be thought of as yet another fundamentally new thing. It is a consequence of the original principle of induction by the following argument. Suppose we know that T is completely inductive. We show every natural number is in T using just the principle of induction. Define a set S as follows:

$k \in S$ if and only if k and every natural number smaller than k is in T.

If we show that S is inductive, by the principle of induction every natural number must be in S, and then it follows trivially that every natural number must be in T. So we show that S is inductive.

Basis step: Before we reached the "etc." above, we argued correctly that every natural number smaller than 0, and 0 itself, must be in T. Then, by definition, $0 \in S$.

Induction step: Suppose that $k \in S$ (induction hypothesis). We show that $k+1 \in S$. Since $k \in S$, by definition of S, k and all smaller numbers are in T. Then all numbers smaller than $k+1$ are in T, which is completely inductive, hence $k+1 \in T$. So $k+1$ and all smaller numbers are in T, hence $k+1 \in S$. End of induction step, and of proof.

By what we just showed, the principle of complete induction is not strictly necessary, since it can be reduced to the ordinary principle of induction. But its use often makes it easier for us to follow a proof that would be less intelligible if ordinary induction were used. We illustrate its use by showing: Every natural number other than 0 and 1 has a prime factorization (if p is prime, it is its own prime factorization). Let T be the set of natural numbers given by:

$k \in T$ if and only if $k = 0$ or $k = 1$ or k has a prime factorization.

We show that T is completely inductive, and then the principle of complete induction gives us the result we want.

Suppose all natural numbers smaller than k are in T (induction hypothesis). We show that k itself is in T. If $k = 0$ or $k = 1$ then $k \in T$ by definition. So now suppose that $k \geq 2$. If k is prime, it is its own prime factorization, hence again $k \in T$. So now also suppose k is not prime. Then k is composite, so we can write $k = i \cdot j$, where neither i nor j is 1. Also, neither i nor j is 0 since $k \neq 0$. Certainly i and j are smaller than k, hence they are in T by the induction hypothesis. Since neither is 0 or 1, both i and j must have prime factorizations, by definition of T. If we combine a prime factorization for i with one for j, we get a prime factorization of k, hence k has a prime factorization, $k \in T$, and the proof is done.

Most commonly we will be using induction to prove things about objects other than numbers. Of course numbers must be brought in somewhere. The following example shows one way of doing this.

Suppose we are considering the family of all words made up of just the letters a and b. If x and y are two such words, we denote by xy the *concatenation* of these words. And we do this similarly for more elaborate cases. For instance, if x is the word ab and y is the word bab, then $xbya$ denotes the word $abbbaba$. We allow the word with no letters, the empty word, which we denote Δ. Now, say we have a set S of words such that

$x \in S$ if and only if $x = \Delta$ or for some $y, z \in S$, $x = ayzb$.

There does exist such a set, and it is unique. Take that for granted for now.

Suppose we want to show that all members of S have some property. To bring in numbers, every word has a *length*. We can use the principle of (complete) induction on these word lengths. Say we want to prove every member of S has as many a's and b's. Well, define a set T of numbers by:

$k \in T$ if and only if every $x \in S$ of length k has as many a's as b's.

If every number is in T, it follows that every word in S has as many a's as b's. Consequently, it is enough to show that T is completely inductive. We do this as follows. Suppose all natural numbers smaller than k are in T (induction hypothesis). We show that k is in T.

Let x be an arbitrary member of S of length k. What we must show is that x has as many a's as b's. Since x was arbitrary, this will establish that $k \in T$. There are two parts to the characterization of S, and so two parts to our argument. If $x = \Delta$ then trivially x has as many a's as b's, and we are done. Otherwise, for some $y, z \in S$, $x = ayzb$. Certainly, y and z are shorter words than x. But the induction hypothesis essentially says that words shorter than x that are in S have as many a's as b's, so this is the case for y and z. It is an obvious consequence that $ayzb$ $(= x)$ has as many a's as b's. And this concludes the proof.

Induction arguments occur often in this book. Except for this appendix we do not use the induction/complete induction terminology. We call both kinds of arguments just proofs by induction.

INDEX